Thorsons Guide to
Amino Acids

D0911598

Thorsons Guide to Amino Acids

Leon Chaitow ND., DO.

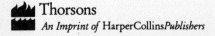

Thorsons
An Imprint of HarperCollins*Publishers*

Thorsons
An Imprint of
HarperCollins*Publishers*
77–85 Fulham Palace Road
Hammersmith, London W6 8JB

Published by Thorsons 1991
Originally published as *Amino Acids in Therapy*
and *The Healing Power of Amino Acids*
10 9 8 7 6

A CIP catalogue record for this book
is available from the British Library

ISBN 0 7225 2492 7

Typeset by Burns & Smith Ltd, Derby

Printed in Great Britain by
HarperCollins Manufacturing, Glasgow

Contents

Introductory Note

In the next few years we will become increasingly familiar with some very strange-sounding substances. The words ornithine and phenylalanine will trip from our tongues just as easily as vitamin E, potassium and selenium now do.

Research is discovering valuable new therapeutic potentials for the 20 amino acids from which we are made. Their healing potential has barely been glimpsed and already (as Chapter 7 shows) some major – and many minor – diseases have been shown to be capable of being modified and improved by simple supplementation techniques.

With effects ranging from pain relief to appetite modification and mood alteration, these simple protein fractions are almost always safe and simple to use, *if the guidelines covering their use are followed* (see Chapter 11). The current scare over the use of tryptophan – very likely the result of contamination of the raw material – shows there is much to learn about securing a sound source of supplements (see Chapter 8).

Before following any self-help advice you are urged to give careful consideration to the nature of health problems you are concerned about, and to consult a suitably qualified health adviser if in any doubt at all. This book is not a substitute for professional advice and treatment, and the author and publisher cannot accept responsibility for any problems arising from experimentation with the methods described.

We strongly urge you to ensure that amino acids purchased for supplementation are from a reputable manufacturer.

CHAPTER 1

About Amino Acids

For health to be at its peak, all essential nutrients are required in their optimum quantities. The optimum quantity of any single nutrient will differ in each individual, depending upon characteristics which are either genetically or environmentally acquired, and which may vary with different conditions such as stress, infection, pregnancy, etc.

The complex interrelationships between the essential nutrients of the body are becoming clear as research unravels the mysteries of biochemical activity. An area of current interest is that of the role of the amino acids. Their relevance to health, and their use in a variety of conditions of physical and mental dysfunction, make them an exciting new therapeutic tool.

However, it is essential that they be used therapeutically (a) only when there is a demonstrable requirement for their administration, and (b) when they are employed in such a way as to ensure that they are suitably combined with those nutrients with which they are normally associated, in their metabolism, and (c) that they are used in such a way as to ensure that no harm comes to the body through the creation of toxicity or of a nutrient imbalance.

These provisos are applicable to the use of all nutrients, whether vitamins, minerals, essential fatty acids or whatever. Because of particular idiosyncracies, an individual may have unusual requirements for a nutrient factor, or combination of factors. Combined with the isolation of particular nutritional needs is the requirement to discover the reasons for that particular need, if indeed that is possible, and to remedy the cause of the problem by suitable lifestyle or dietary or emotional alterations and modifications.

The division of nutrients into groups of particular biochemical activity or origin, such as vitamins, minerals, amino acids, etc. must not blind us to the fact that all the nutrient factors interact. This may break down, or become inefficient, at any point along the chain if any one of the nutrients or their products is not present in its optimum quantity. In all physiological functions there is a level of nutrition relating to each participating nutrient, below which obvious signs of disease will be manifest and above which obvious signs of toxicity will appear. Between the level of obvious deficiency and patent toxicity, lies an area of function in which there are infinitely varying degrees of normality. Many people spend most of their lives at a level just above deficiency – and show some signs of deficiency. Since it is demonstrable that individual needs may vary by large amounts (see next chapter) in the case of the amino acids, the level at which a breakdown of function occurs will differ widely. This makes a nonsense of Recommended Dietary Allowance figures, which at the best can be seen to be applicable to a mythical 'average' human who does not exist. Recommended dietary levels and therapeutic dosages of any nutrient can therefore only give a rough guideline, as to whether the needs of the body are going to be met.

It is self-evident that the intake of any nutrient is but the first step in a chain of events involving the ultimate safe arrival of the nutrient, or its products, at the site of the particular biochemical activity for which it is destined. Whatever stage of a nutrient's journey to its biological appointment is inefficient, or disturbed, may require therapeutic attention, rather than that there should be an automatic assumption that deficiency, on the cellular level, necessarily means that there is a dietary deficiency. Nowhere is this more evident than in amino acid therapy.

Naturopathic methods and modern clinical nutrition are in total accord inasmuch as adequate nutrition is seen to be one of the main essentials for health. The same factors which promote and maintain health are seen to be the factors which will restore it, or allow it to be restored, when health is absent. This is as true for amino acids as for any other nutrient. Unless the correct amount of any of these is present in the correct tissues at the appropriate time there will not exist the possibility for optimal function of the body. The reasons for such a relative local deficiency may be different in one person as against another.

Inherited characteristics may determine the requirement of a particular nutrient in quantities far in excess of what is considered normal. There might be acquired alterations to digestive function, or to the ability of the nutrient to be absorbed. Such acquired absorption and digestive problems are frequently the result of incorrect feeding programmes in infancy, which may produce irreversible damage to these functions.

General nutritional imbalance is also a possible cause of damaged absorption or utilization of nutrients. If the dietary pattern is such as to produce biochemical imbalances or inadequacies of substances which are vital to the processing, transportation or delivery of other nutrients, then that aspect must be dealt with, as a primary consideration. It is also possible that the only factor in a nutrient's inability to find its way to where it should be may be the form in which it is ingested. This can determine its biological availability. The example of orally ingested inorganic iron is clearly such a one. With all these variables, and with the vast and complex interrelation of nutrient factors, one to another, in all their chemical forms as they are swept along in the unending process of the vital life of the body, it is easy to become intellectually overwhelmed.

The ideal to aim for is to come to an understanding of the body's general nutrient and other requirements, and of being able to identify particular needs. Thereafter the realization that the body is invested with self-regulating, self-healing attributes, which can be enhanced by modification of lifestyle, nutritional patterns, stress levels, etc., and which will then operate to improve overall function and restore health, is to have a sound basis for nutritional health. The aim is not to identify nutritional requirements on the basis of 'a particular nutrient for a particular complaint', but to work with the body in its self-regulating efforts (homoeostasis), and to attempt to provide the essential nutrient factors, which may be indicated, as part of a comprehensive approach which takes into account all those factors which might be mitigating against the normal working of the organism. As has been stated, there may be a variety of reasons for the self-same symptoms appearing in different people. It is the causes which require identification, and not the symptoms which require 'curing'.

Amino acids can best be described as the construction blocks from which protein is made. Just as in a child's construction kit the pieces come in different shapes and sizes and yet fit together to make something recognizable, so the more than 20 amino acids each have unique characteristics, and yet are capable of being fitted together into an almost limitless variety of proteins.

Protein is formed by the joining together, into chains, of amino acids and thus far over 100,000 different proteins have been identified in nature, which are the result of variations in the pattern in which the chains are constructed.

The human body alone contains over 50,000 different forms of protein. The total presence of amino acids in the body represents fully three quarters of the body's dry weight (this is excluding the water content). Most of the amino acids in the body can be manufactured out of just eight other amino acids, which are all essential in the diet. This means that our diet has to allow the acquisition of free forms of these eight amino acids for life to continue.

These 'essential' amino acids are critically important to life and health, for out of them the body makes the other amino acids, as well as many of the vital compounds which keep the body working, such as the enzymes, neurotransmitters, mucopolysaccharides etc., not to mention blood, muscles, organs and bones from which we are constructed.

When only a short chain of amino acids is joined together, in a particular sequence, it is called a peptide. When the chain is long, it is called a protein.

The amino acids themselves are constructed from a combination of the following elements: carbon, hydrogen, oxygen, nitrogen and in some cases sulphur.

Every amino acid comes in two forms, a 'left-handed' (L)

and a 'right-handed' (D) form. These two forms are identical in every respect except for the conformation of the subunits of which they are composed. That is to say, although chemically they contain the same elements, in precisely the same quantities and in the same sequence, they are the mirror image of each other, just as the human left hand has the same construction as the human right hand and yet they are different (a right hand cannot wear a left-handed glove for example). Protein chains cannot be formed from a combination of L and D amino acids.

The body is constructed almost without exception from the L forms of amino acid. However, the D forms, which occur in nature, are often found to have therapeutic value and, as we shall see later, the D form of phenylalanine is a particularly valuable asset in treating pain.

The essential amino acids which are required by the adult body (children have slightly different needs, as we shall see) to make the other amino acids as well as the proteins of the body are: L-tryptophan, L-isoleucine, L-lysine, L-threonine, L-leucine, L-methionine, L-phenylalanine, L-valine. Henceforth, we shall drop the 'L' prefix so that it can always be assumed that a named amino acid is the L form. D forms, or a combination of D and L forms, will be clearly described as such. From these raw materials, which are essential elements in the diet, the body synthesizes the other amino acids (nonessential) which are cysteine, cystine, tyrosine, arginine, alanine, glutamic acid, proline, hydroxyproline, glutamine, histidine, aspartic acid, glycine, serine, asparginine, carnitine.

Recent research, however, has questioned the concept of essential and non-essential amino acids. Arginine for example is known to be in short supply in children, and may therefore be considered 'essential' for them, because the young body is incapable of manufacturing adequate amounts from the other essential raw materials, as an adult body can. Histidine is also considered necessary in the diet of infants, whereas it is not considered by all experts to be an essential amino acid for adults.

Furthermore, it is now known that, under certain circumstances, any amino acid can become essential. This is explained in greater detail in the next chapter.

CHAPTER 2

Why We All Have Individual Requirements

In his landmark book on the subject of nutrition, entitled *Biochemical Individuality*[1] Roger Williams Ph.D. establishes the known facts relating to variations in the requirements of essential nutrients. This knowledge is re-emphasized and updated by Professor Jeffrey Bland[2] and the material contained in this chapter owes much to these two researchers.

There is evidence of variations in amino acid excretion patterns, in urine for example, which shows massive differences between individual amino acids such as lysine, where some healthy individuals displayed excretion levels several times that of others. The same individuals were also known to have up to ten times the quantity of particular enzymes present in their saliva. Saliva has been used as a means of determining variations in individual make-up, and the degree of such variation, both as to constituents and as to their individual quantities, is profound.

Williams showed that all individuals tend to maintain a unique, distinctive pattern of amino acid concentration. Studies were conducted to assess human amino acid requirements for the maintenance of nitrogen equilibrium. In a relatively small sample of individuals (about thirty people) it was demonstrated that, in some people, the RDA of amino acids was low by a factor of three. Larger samples of people could be expected to show even greater variations from the norm.

Bland discusses the range of individual amino acid requirements, pointing out that the reported ranges differ by between two- and seven-fold, with an average range of four-fold differences in requirements for particular amino acids. This was in samples of between 15 and 55 subjects. In one small group of college women variations in requirements of amino acids were between three- and nine-fold, with an average in excess of five-fold. Such findings are supported by animal models, and the general conclusion is that RDA of individual amino acids has little practical value, apart from the barest average guideline; and that individual requirements of amino acids, as with all nutrients, are variable to an unpredictable extent. Inborn genetic differences are the main single factor determining these variations. Every stage of any process within the body is capable of being genetically modified to create such variability. It could be that the digestion, absorption, processing into different forms, transportation, storage or utilization of any substance, in any biochemical process, has in some way been modified genetically to produce a unique nutrient requirement pattern. It is this pattern that must be met, within certain limits, in order for health to be enjoyed. It need not be any obvious genetic effect acting directly on the particular nutrient in question, but rather an indirect influence upon it, that creates a need for greater quantities than normal. Any genetic alteration of any of the thousands of enzymes involved with amino acids, in biochemical processes of the body, could result in expressions of imbalance in one nutrient factor or another, including amino acids.

Certainly factors other than the inherited characteristics of the individual can have similar, if less marked, effects. These can include the interaction of those nutrients present, or in short supply, in the diet; stress factors; exercise levels; particular physiological (e.g. pregnancy) or pathological (e.g. fever) states, which also present the body with demands in excess of RDA estimates. Trauma, shock, intense heat and cold, surgery and emotional strain, as well as the use of toxic substances, whether these be in diet (e.g. alcohol, coffee) or in the environment (e.g. heavy metals in water, or atmospheric pollution) or the use of therapeutic or addictive drugs, may all involve increased demands for essential individual nutrients, sometimes by heroic amounts.

The term 'essential nutrient' may also be considered a variable factor in the light of the evidence of biological individuality. An essential nutrient is generally understood to be one which the body is unable to synthesize for itself, and which it is dependent upon the diet to provide. There are some forty-five such substances including, in adults under normal conditions, eight amino acids. However, under certain conditions some of the amino acids, which are usually considered non-essential, in that the body is able to synthesize them, can become 'essential' and require dietary reinforcement in order to maintain health. Such substances are called 'contingent nutrients' by Bland. This phrase has an elegance which encapsulates their ambiguous role. In certain contingencies they become essential, and recognizing the possibility is an invaluable aid in the effort to assess the particular needs of the individual. It appears that in the young and the elderly the chances of nutrients developing a contingent status is greatest. Arginine is synthesized in young people, but not in adequate amounts to meet the needs of the growth period. It is, therefore, at this time a contingent nutrient, and essential in the diet to maintain health.

There are other times when it is quite reasonable to suppose that the ability of the body to produce adequate quantities of particular non-essential amino acids might be wanting. Bland gives the example of histidine which we can produce in more or less adequate amounts under normal conditions. Since histidine is required in the production of histamine there are many opportunities for the body's requirements to outstrip its ability to produce enough. Such increased demands might occur during chronic illness or the use of particular drugs.

All amino acids are potentially contingent nutrients, under suitable conditions, and so the dividing line between the essential, and the non-essential may become blurred. In normal conditions there is a distinction to be made between those amino acids which can, and those which cannot, be self-produced. What is normal will differ with individuals and the stresses imposed upon them. Williams' concept of genetically originating 'diseases of nourishment', i.e.

genetotrophic diseases, is the basis for our understanding of the phenomenon of diversity and individuality as to the needs of each person. The range of augmentation of specific nutrients required to deal with each inborn variation in requirement, in order to normalize, or prevent associated dysfunction and disease, has been estimated by Bland to vary from two-fold, to several hundred-fold, to meet the particular requirements of some people.

At this stage it is only necessary for us to be aware of the certainty that each person possesses variations in needs of all nutrients, and these variations are sometimes extremely marked. It is also pertinent to note a corollary to the above, that there exists a certain possibility that normally non-essential nutrients may be inadequately available, and therefore may be entering the realm of essential substances.

References

1. *Biochemical Individuality*, Roger Williams Ph.D., Texas University Press 1979.
2. *Medical Applications of Clinical Nutrition*, Ed. J. Bland, Keats 1983.

CHAPTER 3

Amino Acids and Protein

By definition an amino acid is any of a large group of organic compounds which represent the end products of protein hydrolysis. Ten of them are considered essential inasmuch as they are required to be present in the diet, at least at some stage of life, when the body is unable to manufacture either adequate amounts, or any at all, for its use. These ten are arginine, histidine, isoleucine, leucine, lysine, methionine, phenylalanine, threonine, tryptophan and valine.

Arginine and histidine are of ambiguous state inasmuch as they may be synthesized by the body, but arginine in young people, during the periods of growth, is required in the diet as well. Histidine is in a similar situation during youth, old age and when degenerative diseases are operating. These two then fall into the contingent category, discussed earlier.[1]

The other essential amino acids, together with a number of non-essential amino acids, such as glutamine and cystine, which have been found to have important therapeutic effects, form the main body of discussion of this book. New amino acids are being discovered, and doubtless therapeutic roles will be ascribed to some of these, and to many of the known but thus far therapeutically non-valuable ones. Our knowledge in this field is in its infancy. However, there are already indications as to the useful application of all of those mentioned above, as well as certain combinations of them.

In considering amino acids in relation to health and ill health there are two main areas to cover. The first looks at particular conditions relating to disorders of amino acid metabolism, resulting in a related pathological state. The second area, and the one which attracts the major interest among nutritionally orientated practitioners, is that involving conditions not specifically related to diseases of amino acid metabolism, and yet which appear to respond positively to dietary manipulation which involves the intake of particular amino acids (and other nutrients).

Such conditions as certain forms of depression; insomnia; herpes infections; weight problems; fat metabolism dysfunction; epilepsy, etc. have all been shown to improve, in suitable cases, by the use of appropriate amino acid therapy. Certain physiological functions have also been enhanced by the selective use of amino acids. These include detoxification of heavy metals; modification of free radical activity; enhanced mental function via neurotransmitter stimulation etc.

The ability of the brain neurones to manufacture and utilize a number of neurotransmitters, such as serotonin, acetylcholine and, it is conjectured, the catecholamines, dopamine and norepinephrine, is dependent upon the concentrations of both the amino acids and choline in the bloodstream. This largely depends upon the food composition at the previous meal.[2] Since the brain is apparently unable to make adequate quantities of amino acids and choline to meet its requirements for neurotransmitter synthesis it is vital that adequate quantities of these precursors are present in the circulation.[3] The role of tryptophan and tyrosine in this process will be considered later. In the current context it is pertinent to simply be aware of the vital role played by amino acids in brain function. It is pointed out that the dry material of the brain comprises more than one third protein,[4] and that stress can create a situation in which non-essential amino acids cannot be adequately produced to meet its needs.[5] A number of researchers have shown that such a situation can result in a range of mento-emotional symptoms, such as depression, apathy, irritability, etc. The subsequent imbalance in uric acid levels resulting from incomplete amino acid synthesis, and consequent utilization of free amino acids as fuel by the body, can result in self mutilating behaviour in children.[6]

Neurotransmitters can be seen as the chemical link whereby one neuron, or a group of these, communicates with another. The importance of the neurotransmitters that are directly related to the nutritional intake of particular foods is self-evident.

Some of the vital consequences of inadequate amino acid synthesis are therefore of potentially dramatic import in many current social and medical diseases. Before moving on to look at pathological states resulting from amino acid metabolism defects, it is important that we examine the basic role of protein and the relationship of the amino acids with it.

Protein

Life without protein is not possible. Growth development and function depend upon it, and it in turn depends upon the correct supply of amino acids. Apart from water, the next most profuse substance in the body is the amino acid group. The matrix into which these substances are incorporated is protein. The structure of all amino acids is similar in that a carbon atom, and an amino group (containing nitrogen), and a carboxyl group are always present. Those amino acids already present, by virtue of being synthesized in the body, are known as Non-Essential Amino Acids (NEAA) and the others, which must be derived from the diet, as Essential Amino Acids (EAA). Both groups are required, in order for protein synthesis to be completed satisfactorily. If one of the EAA is absent or inadequately supplied, then protein synthesis will not be possible. All the essential amino acids must therefore be present in the digestive tract at the same time. Proteins in foods differ substantially in their composition of amino acids. Those that contain all the EAA are termed complete proteins, and those that do not are called incomplete proteins (vegetable sources). Incomplete proteins can become complete by the judicious combining of appropriate vegetable sources, such as grains and pulses (ratio of 2:1).

Total protein requirements will vary with age, sex, body-type, occupation, stress levels, exercise pattern etc. Protein is required for the formation and maintenance of blood, muscle, skin and bone as well as the constituents of blood such as antibodies, red and white cells etc. Hormones, enzymes and nucleoproteins are all dependent upon protein. Certain racial groups are better able to metabolize the protein in their food than others. For example, Orientals can survive in good health on a lower protein intake than Caucasians of the same age, sex and body-type.

Most amino acids can be converted into other amino acids, thus methionine can be altered to form cysteine; and tyrosine can be formed from phenylalanine. If protein is not able to be synthesized, due to an inadequate presence of EAA, then the body can utilize the remaining amino acids as fuel. However, since the body cannot oxidize the nitrogen portion of the amino acid, there is a degree of residue from such a process. This residue joins the breakdown products of protein in the body as urea or uric acid.

Despite earlier prejudice against the use of incomplete protein sources by vegetarians, it is now acknowledged that this mode of eating provides all that is required for a healthy body, as long as combinations of vegetable protein sources are adequate.

The relative quantities of amino acids contained in any particular food determine its nutritional value. Whilst individual amino acids may be absent from some vegetable sources of protein, there is nothing 'inferior' about the ultimate protein produced when correct vegetable protein combinations are simultaneously introduced into the eating pattern, so as to allow the amino acid 'pool' to contain its correct complement for protein synthesis. Wheat products, which are deficient in the essential amino acid lysine, are moderately endowed with methionine, whereas the reverse is true for legumes. If both were eaten at the same meal the proportions of lysine and methionine would then complement each other.[7]

Absorption of dietary amino acids, for protein synthesis by the liver, occurs from the intestines. Such synthesis is impaired if the EAA tryptophan is not consumed at the same meal as the other amino acids.[7] One consequence of protein inadequacy is termed 'negative nitrogen balance'. This occurs when the intake and synthesis of protein fails to meet the overall level of total nitrogen loss via sweat, urine, faeces etc.

It is calculated that whilst 32 per cent of total estimated protein requirement in children should be supplied as essential amino acids, this level drops to only 15 per cent in adults. On a body weight basis adults therefore require 78 mg of EAA per kilogram of body weight, whereas children require 214 mg of EAA per kilogram of body weight per day. As has been stated, other variables in determining protein requirements may include stress, infection and heat, which can all cause increased nitrogen loss. Increases in muscle mass as a result of intensive exercise or heavy work will also call for increased protein synthesis, and therefore of greater amino acid intake. It is important to realize that calculations of protein requirements are only valid if the body's energy requirements have been met. For if energy intake is not adequate some dietary and/or tissue protein will be oxidized or converted into glucose in the liver to meet energy needs. Efficiency of nitrogen utilization is dependent upon the total calorie intake, and this is in turn a factor in deciding protein requirements.

Studies of nitrogen balance have produced estimates of RDA of protein for different groups. These do not, of course, take into account individual inherited or acquired variables as to requirements for particular amino acids. Whilst standard RDA levels seem to indicate that most people in industrialized countries obtain adequate protein levels, the number of variables (age, sex, occupation, health status, racial group, stress levels etc.) as well as the fact of biological individuality, makes these of questionable value. It is evident from surveys of food intake amongst urban teenagers for example that inadequate protein intake is not uncommon.

In America the intake of protein in healthy adults is estimated to be the equivalent of 90–100 g per day, which represents between 15 and 17 per cent of the total caloric intake. Minimum protein requirements in an adult are set at 35–40 g per day, and the RDA is put at 44–56 g per day (in a healthy adult). The excess of protein over actual requirements indicates a more than adequate protein intake according to many nutritionalists. How then can there be the possibility of a deficiency in amino acids? The overall imbalance in nutrient intake, as well as the genetically determined variables in requirement, together create a situation in which particular needs may not be met, even in the face of the veritable deluge of protein. As health levels decline in the face of dietary patterns which bear little relation to the human body's actual needs, and as this factor, together with such elements as stress and pollution further mitigate against normal function, so there develops the possibility of conditions such as pancreatic insufficiency. This phenomenon is of vital importance to our understanding of amino acid deficits.

It has been pointed out that the pancreas is faced with the task of making useful by-products from ingested food and chemicals, as well as buffering against reactions to foods and chemicals.[8] For a variety of reasons the pancreas may be overstimulated, and one such reason is the very fact of excessive protein intake. The proteolytic enzyme production capacity of the pancreas can, like any other function, become impaired through over-use. This is especially true in an organ like the pancreas with multi-purpose functions, all of which may be being overtaxed simultaneously. A variety of factors can cause the pancreas to lose efficiency. These include the assault on it by the monumental amounts of sugar which it is obliged to handle via its insulin production. Evolutionary adjustment to the increase in sugar consumption cannot occur in the short space of time involved in this dramatic change in human nutritional habits. Alcohol also has a direct ability to induce pancreatic insufficiency, as have a wide range of drugs, coffee and cigarettes.[9] Excessive fat intake is a further contributory cause. The first effects of such a pancreatic insufficiency are a reduction in bicarbonate production, leading to symptoms which are frequently dismissed as gastritis. This is followed by reduced enzyme activity and finally insulin production is affected.

Inactivation of, or insufficiency in the production of, proteolytic enzymes, from the pancreas, such as trypsin, chymotrypsin and carboxpeptidase, can result in poor digestion of proteins into amino acids. A further likelihood is that protein molecules might be absorbed in their undigested forms, which can provoke inflammatory reactions, sometimes in distant tissues and organs. If at the same time the circulating anti-flammatory enzymes are deficient, as a further consequence of pancreatic exhaustion, then the ability of the body to deal with such inflammatory reactions (allergic or otherwise) will be reduced or absent. The ability of pancreatic insufficiency to interfere with amino acid digestion is, however, our main concern. Should inadequate breakdown of ingested proteins take place, and amino acid deficiency result, despite high levels of first class protein in the diet, the consequences could include difficulty, or inability, on the part of the body to produce adequate enzymes, hormones, antibodies and new tissues. The likelihood would then also exist for excessive demands to be made on a wide range of minerals and vitamins, particularly pyridoxine, zinc and magnesium, leading to deficiencies in these. The immune system's ability to adequately defend the body under such conditions would be severely compromised.

We have, therefore, a picture in which the very presence of excessive protein in the diet (a fact of life in many western cultures) is a contributory cause of the deficiency of adequate protein levels within the system due to pancreatic insufficiency. The consequences of proteolytic enzyme deficiency resulting from pancreatic insufficiency, which itself results, in part, from specific amino acid deficiency (methionine etc.), is most important in our understanding of the role of diet in the production of amino acid imbalances and disease such as allergy. Amino acid imbalances acquired via environmental and dietary sources, superimposed upon those acquired by genetical idiosyncracies, thus create requirements of individual nutrients in excess of average, and this is the overall justification for utilizing amino acids and other nutrients therapeutically in the manner discussed in this book.

References

1. *Medical Applications of Clinical Nutrition*, Ed. J. Bland, Keats 1983.
2. 'Nutrients and Neurotransmitters', *Contemporary Nutrition*, Vol.4 No. 12, 1979.
3. *Archives of Pharmacology*, 303:157-64. 1978.
4. *Orthomolecular Psychiatry*, Vol.4, No.4. pp297-313, 1975.
5. *Journal of Clinical Nutrition*, 1:232, 1953.
6. *Schizophrenia*, 1:3:1967.
7. *Contemporary Nutrition*, Vol.5, No.1, 1980.
8. *A Physician's Handbook on Orthomolecular Medicine*, R. Williams, D. Kalita, Keats, 1979.

CHAPTER 4

Amino Acids, the Body Cycles and the Nervous System

The biochemistry of the body is intensely complex, and in order to come to terms with amino acid therapy it is necessary to have a basic understanding of some of the major processes in which they are involved or which affect them. Two such cycles of activity are the urea cycle and the citric acid cycle (Krebs cycle).

The major toxic byproduct of amino acid activity in the body is ammonia, and in order to prevent it from reaching harmful levels in the system the body undertakes a sequence of metabolic reactions, which turns the unwanted nitrogenous wastes into urea for subsequent elimination via the kidneys. The liver is the main site of this activity. This is called the urea cycle. Were ammonia to be allowed to reach toxic levels a number of serious consequences would occur.

The body produces an 'energy carrier' called adenosine triphosphate (ATP) which is involved in many metabolic processes concerning carbohydrates and amino acids. The production of ATP is the result of activity in what is called the citric acid cycle, or Krebs cycle, in which chemical respiration and oxidative phosphorylation produce carbon dioxide and bound hydrogen atoms. This leads to an electro-transfer reaction which results in ATP. One other product of the sequence of metabolic reactions in the citric cycle is the formation of alpha-ketoglurate, which is the primary amino acid receptor. Acting with vitamin B_6, in the form of the coenzyme pyridoxal phosphate, alpha-ketoglurate detaches the NH_2 molecule from dietary protein. Thus it counteracts excess acidity. This cycle of natural combustion of nutrients (citric acid/Krebs cycle) can be severely interfered with by the presence of ammonia, the breakdown product of amino acid activity. By interfering with and depleting the levels of alpha-ketogluterate, ammonia produces a toxic effect which can lead to a wide range of symptoms, such as: irritability; tiredness; headache; allergic food reactions (especially to protein foods); and also, at times, diarrhoea and nausea. It is also possible for mental symptoms to manifest, including a confused state. Alpha-ketoglutaric acid is the precursor of glutamic acid, the principal amino acid contributor to brain energy supplies. In the conversion to glutamic acid from alpha-ketoglutaric acid, other amino acids are metabolized by the transaminase enzyme, and pyridoxal phosphate (B_6). When such an exchange is interfered with, for one of a number of reasons, there occurs an amino aciduria. Among the reasons put forward for the development of such a situation are: Vitamin B_6 deficiency; zinc deficiency, resulting in an inability to transform B_6 into pyridoxal phosphate; inadequate alpha-ketogluterase, etc.

It is possible for alpha-ketogluteric acid to be deficient when excessive ammonia is present, and also if the citric acid cycle is interfered with, or if manganese assimilation is not adequate.

Alpha-ketoglutaric acid may be usefully supplemented, in the diet, in cases where an excessive amount has built up, or where there is evidence of impaired citric acid cycle function. It may also be necessary when amino acid transfer is diagnosed as inadequate, and manganese is simultaneously found to be deficient. The therapeutic dose of alpha-ketoglutaric acid is between 500 mg and 2500 mg daily, together with pyridoxal phosphate (B_6), and a low protein diet.[1] Interruptions in the primary mechanism of nitrogen waste disposal, the urea cycle, can result in a variety of enzymatic deficiencies. It has been found that arginine can positively modulate certain aspects of such interruptions.[2]

Philpott maintains that there is evidence from amino acid profiles that alpha-ketoglutaric acid is the most deficient substance that can be demonstrated in cases of either physical or mental degenerative disease. He sees its involvement in a number of enzyme steps, associated with vitamin

Weakness

B_6, as well as its role as a precursor of glutamic acid, as being profoundly influential in the production of symptoms when it is deficient. The first such symptom to be noted being weakness. He points to the link between the citric acid cycle (energy generation sequence) and the urea cycle (nitrogenous waste disposal sequence) as being aspartic acid. The improvement in alpha-ketoglutaric acid status that might be achieved by supplementing its citric acid precursor, improves citric acid cycle function. Supplementation of aspartic acid would be expected to have a similar effect on the urea cycle, resulting in ammonia detoxification. Philpott bases his comments on the evidence of a large number of amino acid profiles, in cases of physical and mental degenerative disease. He states that the approach of utilizing citric acid and aspartic acid supplementation will more often than not be the correct one in such cases. If, however, reliable amino acid profile testing were available, specific evidence would then be to hand for confirming the requirement for such supplementation.[3] This is obviously more desirable than an arbitrary assumption.

Levine indicates a further ramification which involves the effects of stress on amino acid status.[4] In normal aerobic metabolism thirty-eight molecules of ATP (energy carrier) are produced for each molecule of glucose metabolized. In states of shock, oxygen consumption and supply decreases and acidosis ensues. This results in as little as two molecules of ATP being produced from each molecule of glucose. Low ATP precludes the biosynthesis of protein, and the derangement of amino acid metabolism which follows can result in many complications. Under flight simulation stress it has also been shown that there was raised excretion of basic and neutral amino acids concurrent with a lowered level of acidic amino acid excretion. The result of this for any length of time is the production of an acid state. Whilst dysfunction of the urea cycle may be the result of multiple enzyme deficiencies, it is frequently the result of impairment of the enzyme arginase. This would be indicated by high levels of arginine in the urine. Other defects in the urea cycle might be indicated by excessive amounts of ornithine, or citrulline. Arginase deficiency is usually accompanied by hyperammonaemia, with glutamine levels also elevated. Symptoms would usually relate to the effects of ammonia accumulation on carbohydrate metabolism, and upon the effect on neurotransmitters. Headache, motor problems, hyperactivity, irritability, tremors, ataxia, vomiting, liver enlargement, and even psychosis may occur. A requisite cofactor of arginase is manganese, and deficiency of this can result in increased excretion of arginine via the urine. Lysine and ornithine are inhibitors of arginase, and a diet high in lysine (such as that suitable in herpes infection) may be indicated in such defects of the urea cycle as well as supplementation of essential amino acids (including tyrosine and cystine).[5] A similar pattern of high arginine excretion (together with ornithine, cystine and lysine) may occur in the amino acid transport disorder cystinuria. This possibility can be excluded by determination of plasma arginine, and blood ammonia levels.

Amino acids and neurotransmitters

Individual types of amino acids have particular characteristics. Some are capable of influencing body processes because they are essential to the formation of neurotransmitters, substances which are used in the brain and by the nervous system to increase or decrease the efficiency and rapidity of nerve transmission. The ability of the brain to receive and to transmit messages depends upon these neurotransmitters, which are themselves dependent upon particular amino acids. All functions of the body depend upon sound nervous interconnection. This allows organs and muscles to report back to the higher centres as to their status, and for receiving instructions from the higher centres, as to their behaviour and needs. The coordination and regulation of all the millions of messages that are constantly going on in the body, depend upon neurotransmitters and therefore on amino acids. Amino acids are especially important where nerves interact (synapse), where information is passed on and received. Some of the neurotransmitters have a stimulating, excitatory function and others have a calming, inhibitory function.

The scope and use of appropriate amino acids in therapy can therefore be seen to be enormous. Unless all the amino acids, in their free form, are present in adequate amounts, there will be imbalances in the neurotransmitter function, and a variety of nervous and emotional problems will result. The very energy of the brain is dependent upon certain amino acids. The two amino acids used as examples of the value of this class of nutrients in Chapter 1, tryptophan and phenylalanine, are both of profound importance in their relation to brain and nerve function, as we shall see.

Another major area of activity of some of the amino acids is as detoxifiers of the body. The sulphur rich amino acids (methionine, cysteine, cystine) are especially capable of this sometimes life-saving task. These have the ability to chelate (lock onto) heavy metals such as lead, mercury and aluminium, which are toxic to the body, and to actually remove them from the system.

They are also capable of damping down damaging processes in the body relating to oxidation of certain substances such as fats. When toxic substances are present in tissue or in the bloodstream, there is potential for what is called free radical damage, as fractions of the oxidizing substance cascade around the area creating tissue damage. These processes which are thought to result in such cell changes as occur in arteries before they become atherosclerotic, and to cells before they become cancerous, are controlled by free radical scavengers or quenchers, of which the sulphur rich amino acids are a major part. Vitamins A, C, and E and the mineral selenium are also anti-oxidants which reduce free radical damage.

In the next chapters we shall look at the roles which have been defined for the amino acids as well as their major therapeutic effects.

Future research will doubtless open up new avenues of therapeutic potential, as it will also most certainly discover new amino acids. Several of these have been noted in the past few years such as γ-carboxyglutamic acid and β-carboxyaspartic acid. What we have at present is a working knowledge of the major amino acids, with a fair idea of how to use these therapeutically. Protein power is now available for us to use in the quest for better health.

References

1. Pangborn, Jon, Bionostics Inc/Klair Laboratory, Pamphlet.
2. Jay Stein, (editor), *Internal Medicine*, Little Brown, 1983.
3. Philpott, W., Philpott Medical Center, Oklahoma City, Pamphlet.
4. Levine, Stephen. Allergy Research Group, Pamphlet.
5. Stanbury, J. et al. *The Metabolic Basis of Inherited Diseases*, McGraw-Hill, 1983.

CHAPTER 5

Disorders of Amino Acid Metabolism

There are a recognized number of disorders relating to amino acid metabolism. It is also generally acknowledged, by even the most conservative of medical experts, that it is reasonable to assume that new inborn errors of amino acid metabolism will continue to be discovered and described, as the overall knowledge of amino acid metabolism develops. Disorders of amino acid metabolism, transport and storage, currently recognized by orthodox medicine do not individually involve large numbers of the population, although their overall combined incidence is substantial. Those diseases which result from amino acid metabolism defects, as a rule, affect mental faculties and result in a reduced life expectancy. Those that involve disorders of transportation and storage of amino acids are associated with a wide range of symptoms. Diagnosis of these disorders requires access to skilled clinical laboratory facilities.

If we are aware of the sort of problems which arise when there is a complete absence of particular amino acids it helps us to understand the sort of symptoms which might arise when there is partial absence, or deficiency. Such deficiencies are far more likely to occur – and to require medical attention – if there exists either an increased requirement for the amino acid which is genetically predetermined, or if there exist digestive weaknesses, relating perhaps to pancreatic insufficiencies, as discussed in Chapter 3.

Diseases directly related to amino acid metabolism problems

Glycine	Nonketotic hyperglycinaemia
	Ketotic hyperglycinaemia
Alanine	Lactic acidosis
Valine	Hypervalinaemia
	Maple Syrup Urine Disease (MSUD)
Isoleucine	Propionic acidaemia
	MSUD
Leucine	Isovaleric acidaemia
	MSUD
Methionine	Hypermethioninaemia
Cystine	Cystinosis
	Cystinuria
Serine	Hyperoxaluria II
Threonine	Hyperthreoninaemia
Phenylalanine	Phenylketonuria (PKU)
	Atypical PKU
Tyrosine	Hereditary tyrosinaemia
Tryptophan	Tryptophanuria
Proline	Hyperprolineaemia I & II
Glutamic acid	Pyroglutamic acidaemia
Histidine	Histidinaemia
Arginine	Hyperargininaemia
Lysine	Hyperlysinaemia
Argininosuccinic acid	Arginosuccinicaduria
Ornithine	Hyperornithinaemia
	Ornithine aminotransferase deficiency
Citrulline	Citrullinaemia
Homocystine	Homocystineuria
Pipecolic acid	Hyperpipecolicaemia
	Zellwagers' syndrome
b-Alanine	Beta-alaninaemia

Some amino acid diseases

Phenylketonuria (PKU): This genetically inherited condition is one of the best known and most studied of the amino acid metabolism problems, with an incidence of one person in 14,000 in the United States. It is the result of an enzyme deficiency (hepatic phenylalanine hydroxilase) which prevents phenylalanine from being converted into the amino acid tyrosine. Symptoms can include severe mental retardation, lack of skin and hair pigmentation, eczema and rashes, fits and abnormalities of the brain, as well as microcephaly (abnormally small head).

The brain dysfunction symptoms seem to relate either to excessive (toxic) build-up of phenylalanine or reduced levels of tyrosine, which result in inadequate neurotransmitter activity. A neurotransmitter is a substance which chemically enhances or retards nerve transmission depending upon its type. When neural activity in the brain either speeds up or slows down, moods of elation and excitement or lethargy and depression result.

If dietary treatment is introduced to cases of PKU in the first month, an IQ of around 100 can be achieved (below average but allowing a reasonable quality of life). If untreated, PKU results in an IQ of under 50 (usually around 20) which means that severe retardation is observed. The dietary treatment involves maintaining a low intake of phenylalanine in the food (averaging 250 to 500 mg daily) until age 8 at least.

Histidinaemia: This condition is the result of a deficiency of the enzyme histidine-a-deaminase, which would normally convert histidine into urocanic acid. Once again symptoms affect brain function, leading to retardation in around half of all cases, as well as speech defects. This condition does not respond to dietary manipulation, even if histidine levels are kept low.

Urea cycle disorders: This is the important body cycle which deals with detoxification of the by-products of protein metabolism such as ammonia and aspartic acid, processing them into urea. If this cycle is disturbed – and there are many different causes of disturbance – a build-up of ammonia in the blood occurs, leading to mental retardation, protein intolerance (allergy) seizure and sometimes coma or death (if untreated).

Transfusion and dialysis are required for initial treatment followed by a very low protein diet and manipulation of amino acid status by specific supplementation (including arginine).

Branched Chain Amino Acid Diseases: These involve all or some of the amino acids leucine, isoleucine or valine. The best known of the problems involving this group of amino acids is the odd-sounding, *maple syrup urine disease.* MSUD results from a deficiency of the enzyme keto acid carboxylase, which is involved in the breakdown of these three-branched chain amino acids. Deficiency results in accumulation of the amino acids (mainly leucine) and a sweet smelling urine – hence the name of the condition.

In new-born infants the condition is characterized by vomiting, lethargy, seizure (fits), increased muscle tone (mild spasticity) and sometimes death. With early dietary strategies, which limit the intake of these amino acids from food, patients respond well. When the condition occurs later in life, stress factors, infection and excessive protein intake have all been implicated.

Various forms of branched chain amino acid disease respond to treatment using Vitamin B_{12} and Biotin (a B vitamin), as well as strict dietary control.

Homocystinuria: This is a condition which is again the result of an enzyme deficiency (cystathionine synthetase) without which the amino acid methionine cannot be altered to cystathionine, resulting in symptoms such as a failure to thrive in infants, the presence of a very light complexion as well as mental retardation. In some cases life-threatening circulatory disabilities involving veins and arteries occur. Treatment using a diet low in methionine, and high in the amino acid cystine, as well (in some cases) supplementation with high doses of vitamin B_6 (pyridoxine) are helpful in most cases. (See also notes of Methionine on page 49).

Retinal Gyrate Atrophy: This condition involves progressive degeneration of the retina and other parts of the eye structure and commences with night-blindness, progressing to loss of peripheral vision and ultimately blindness by middle age. It is caused by a congenital deficiency of the enzyme ornathine-a-aminotransferase. A diet which is based on foods low in the amino acid arginine (as used in anti-herpes strategies, see notes on Arginine and Lysine on pages 37 and 46 and Herpes on page 120) can stabilize the function of the eyes.

Hartnup Disease: Unlike most of the conditions described, this disease results from a problem of transportation involving the amino acids alanine, serine, threonine, valine, leucine, isoleucine, phenylalanine, tyrosine, histidine and tryptamine. The presence of these amino acids in the urine can rise by up to 1000 per cent.

The symptoms include an eczema-like rash on the arms, legs and face which is very sensitive to light; a tremor on movement, visual problems, hallucinations and mental retardation. The symptoms are seldom constant but come in bouts, often related to stress, infection, exposure to sunlight and treatment with sulphur-based drugs. The condition is similar to chronic vitamin B_3 (nicotinamide) deficiency (pellagra) to the extent that it improves (or at least the skin and neurological symptoms do) when this vitamin is supplemented.

When the amino acid status of normal children is compared with that found in children who are either mentally handicapped or physically handicapped, we can see quite marked variations. One study of this sort which was carried out in Manchester (*The Lancet* 11, 10–14, 1981) involved 75 physically or mentally handicapped children (these included children who were quadriplegic, spastic, epileptic, diplegic, or who had Down's Syndrome, psychiatric disorders, mental or developmental retardation or congenital cataracts) all of whom had the amino acids in their urine examined and compared with the findings from 59 normal children.

The results showed very high levels of glycine, taurine and cystathionine in the handicapped group with a few also showing high levels of phenylalanine, serine, tyrosine, histidine and asparagine. The researchers believe that such tests, conducted after a high protein intake, should be carried out on all children who fail to develop normally so that strategies can be developed to bring the biochemical dysfunctions into normal ranges.

Long-term care could involve careful dietary patterns once the underlying biochemical imbalances are identified and understood. Such approaches seem effective in many of the 'simple' amino acid problems described above and could be a useful approach when used in the complex imbalances seen in this type of child.

We should keep in mind that some of the conditions discussed in this chapter are genetically induced, while others are influenced by environmental factors such as stress and infection. The latter would probably be helped by a more wide-ranging approach taking in general health factors. Use of amino acids and other nutrients should play just one part in a more general, holistic, treatment strategy.

The conditions outlined above are only a sample of the many and varied disorders which involve, wholly or partially, essential or non-essential amino acids. In the next chapter we will be looking at details of individual or combined amino acids which are now being used extensively by nutritional therapists.

We will look at these individually, with brief indications of the sort of conditions in which they are found to be helpful, along with some guidelines as to which nutritional cofactors are essential or which often work best with them. In some instances the specific foods in which they are most likely to be found will be mentioned, otherwise supplementation is the only satisfactory strategy available.

It should be clear by now that amino acids are at least as powerful and vital in their potential for good as the more glamorous vitamins and minerals with which we have now become so familiar. They are, of course, equally powerful in their potential for harm when deficient, or present in excessive (toxic) amounts.

References

Scriver, C.R. and Rosenberg, L.E. *Amino Acid Metabolism and its Disorders*, Saunders, (Philadelphia), 1973.

Stein, J.H., (Editor), *Internal Medicine*, Little, Brown and Co. (Boston), 1983.

CHAPTER 6

Individual Amino Acids: Therapeutic Roles

Essential amino acids (EAA)
Arginine
Histidine

These two amino acids are essential in the growth period of life and sometimes in adult life through acquired, or genetic, factors.

Isoleucine
Leucine
Lysine
Methionine
Phenylalanine
Threonine
Tryptophan
Valine

Non-essential amino acids (NEAA) with therapeutic characteristics
Proline
Taurine
Carnitine
Tyrosine
Glutamine and Glutamic acid
Cysteine and cystine
Glycine
Alanine
b-Alanine
Gamma Aminobutyric acid (GABA)
Asparagine and Aspartic acid

Citrulline
Ornithine
Serine
Glutathione (Cysteine, glutamic acid, glycine)

Arginine

This is an essential amino acid (EAA) only during the period of growth, since the body can manufacture it later in life. It can itself be turned into the amino acid ornithine (and urea) which makes it important in certain of the detoxification processes which take place in the liver (the urea cycle). Many important body tissues such as collagen and elastin – as well as substances such as insulin and haemoglobin – require arginine for their manufacture. Eighty per cent of seminal fluid comprises arginine.

Professor Roger Williams[1] reports that while arginine is not an EAA in most adults, in some people it may be required in the diet. He therefore calls it a 'contingent' nutrient, since it becomes essential in the diet under certain circumstances. For example when certain local changes result in a lack of sperm (not where this is the result of a more general disease) the condition is called idiopathic hypospermia. Williams reports this has been successfully treated using 8 g of arginine daily. He says, 'It seems reasonable to suppose that certain individuals would be found who would have partial genetic blocks which would make the production of arginine from other amino acids difficult.' Such an individual might develop idiopathic hypospermia and would require arginine 'contingently' in order to produce sperm.

Borrmann[2] reports that arginine is useful in cases of sterility and that it acts as a detoxifying agent.

Anyone with herpes simplex infections should avoid a high intake of arginine according to many authorities.[3,4,5] Foods rich in arginine should therefore be avoided by anyone with such viral infections (see Lysine section page 46 for dietary indications in such cases, as well as Herpes section page 120). Arginine is glycogenic (it can be turned into sugar) and is involved in what is probably the most important part of the urea cycle, the process during which ornithine and urea are formed from ammonia, thus freeing the body of this dangerous breakdown product of protein metabolism.

The American nutritional researcher William Philpott[6] tells us that arginine enhances immune function and that although we can starve the herpes virus of arginine and provide it with lysine instead (thus slowing or stopping its activities) we can in fact harm the body in this same way. Long-term imbalances would harm the immune system, he says, as well as harming the way we handle carbohydrates. Philpott also suggests that arginine can usefully be linked ('chelated') with minerals such as manganese, to make an easier access to the body.

Many seriously ill or injured people are now fed by drip feed (parenteral nutrition) with various mixtures of nutrients being included in the liquid food used. In animal experiments performed in order to help improve such formulations it has been observed that additional arginine has the effect of improving post-wound weight loss as well as accelerating wound healing.

The important thymus gland, which is a vital part of immune function, is seen both to enlarge and become more active with arginine supplementation in animals. Tests were conducted in order to see whether this was a result of the arginine increasing pituitary activity (arginine is known to increase secretion of growth hormone from this major gland). In fact this was confirmed since it was only when there was a healthy pituitary function that arginine was able beneficially to improve thymus function (and wound healing).[7] The researchers concluded, 'We suggest that supplemental arginine may provide a safe nutritional means to improve wound healing and thymic function in injured and stressed humans.'

Other studies (using rabbits) showed arginine to be helpful in reducing high levels of cholesterol in the blood and in helping the treatment of atherosclerosis (hardening of the arteries).[8]

When female rats were deprived of adequate arginine sexual maturity was delayed.[9] In this particular study rats given a diet containing 56 per cent normal arginine reached puberty at the correct time, but their ovaries were small and the ovulation rate was low compared with rats on normal or high levels of arginine. The implications for non-fertile women of such studies remain to be discovered.

When arginine deficiency exists in experimental animals a wide array of altered functions are seen. The ability to handle glucose (sugar) is diminished, as is insulin production; the liver's ability to process fats is also affected, resulting in accumulations of fatty deposits. Again we cannot yet transfer these findings to humans, and research continues.

When a diet low in arginine is followed (in an anti-herpes strategy for example) a danger clearly arises: if these animal studies are a correct predictor of what would happen in humans, that a fatty build-up might take place in the liver. Strategies to counteract such a possibility were examined.[10] It was found that the simple means of substantially increasing dietary fibre intake (using guar gum rather than wheat bran) increased nitrogen elimination through the bowel, and as a result the need for excretion of nitrogen through the kidneys (and on into urine) was decreased sufficiently to reduce any possible side effects of arginine deficiency.

This is a useful method, which people who are following an 'anti-herpes' diet or taking lysine supplements while reducing arginine could adopt in order to avoid the risk of the lowered immune function which Philpott predicts with such an unbalanced amino acid intake.

Incidentally it is known that diabetics benefit from the use of guar gum as it improves insulin function, allowing better tolerance of glucose. A recent report[11] described what happened to a child congenitally deficient in an enzyme (carbamyl-phosphate synthetase) which is involved in arginine metabolism. The child stopped growing and developed a severe rash; these symptoms vanished when 400 mg per day of arginine was supplemented. When the arginine was stopped experimentally, symptoms recurred. It was then found that it was necessary to supplement with 800 mg arginine daily to normalize function and keep growth rates normal. This is an example of what Dr Bland calls a contingent state (see page 12) in which arginine supplementation became essential.

But what happens when there is too much arginine in the system? Excessive arginine was found on examination of a group of patients suffering from periodic catatonic states. (These are a type of schizophrenic symptom in which the individual seems to lose the will to talk or move, standing and sitting in one position for long periods, and resists all active movement or speech, sometimes punctuated by short outbursts of violent speech or activity). These patients were also shown to have excessive levels of the amino acid glutamine. Although it was not known whether the excessive amounts of arginine and glutamine were a cause of the problem, they were certainly part of it.[12]

Arginine is found in rich supply in the following foods: peanuts, peanut butter, cashew nuts, pecan nuts, almonds, chocolate and edible seeds. It is present in moderate quantities in peas and non-toasted cereals as well as in garlic and ginseng.

A recent controversial application for arginine has been promoted by a number of American researchers. Basing their recommendation on arginine's known ability to promote growth hormone production, authors Durk Pearson and Sandy Shaw[13] and Earl Mindell[14] suggest that weight reduction, and muscle building, can be enhanced by its supplementation. In Pearson's view, ornithine is also called for in this regard. Mindell states: 'Stimulation of growth hormone in the adult benefits an improved immune response, allowing our bodies to repair themselves more efficiently. In the process, the release of extra amounts of growth hormones in adults can lead to the metabolism of stored fat and the building and toning up of muscle tissue.' The dosages suggested in this particular programme are 2 g, on an empty stomach before retiring, and 2 g on an empty stomach one hour prior to vigorous exercise. Mindell warns of adverse effects after several weeks of this programme in mature adults, where the first side effects noted are reversible thickening and coarsening of the skin. It should be emphasized that there are no long-term studies in this area of massive supplementation of amino acids and the author of this work reports the above but does not add his voice in support of anything but the short term use of such dosages. (See also Ornithine, page 83).

It is reported[13] that schizophrenics should be cautious in their use of arginine as it may result in aggravation of symptoms as a consequence of promotion of growth hormone release. Doses of over 30 mg daily of arginine in anyone who has a history of schizophrenia, is therefore not recommended.

References

1. Williams, R., *Biochemical Individuality*, University of Texas Press, 1979.
2. Borrmann, W., *Comprehensive Answers to Nutrition*. New Horizons, Chicago, 1979.
3. Passwater, R., *Energy Medicine* Vol II. No.1-11, 1980.
4. Kagan, C., *Lancet*, 26 Jan 1974.
5. *Dermatologica* 156:257-67 (1978).
6. Philpott, W., *Manganese-Arginine Complex*, Klaire Laboratories leaflet.
7. *American Journal of Clinical Nutrition*, 37(5) p786, 1983.
8. *Atheroschlerosis*, 43, 1982 p381.
9. *Hormone and Metabolic Research*, (1982) 14(2) pp471-5.
10. *Journal of Nutrition*, 113(1) 131-7, 1983.
11. *Am. J. of Diseases of Children* 135(5) 437-42, 1981.
12. *Journal of Mental Science*, 104 No.434 pp188-200, Jan 1958.
13. Pearson and Shaw, *Life Extension*, Warner Books, 1982.
14. Mindell, Earl, Ph.D., *Arginine*, pamphlet, 1983.

Histidine

This is regarded as an EAA in the growth period, but, since healthy adults are shown to be capable of synthesizing amounts adequate to their needs, it is termed a NEAA in adult life. The neurotransmitter histamine is derived from histidine and, as Hoffer puts it:[1] 'It is not difficult to believe that histidine levels will influence histamine levels.' When the acid group is removed from histidine it becomes histamine. Both histamine and histidine will chelate with trace elements such as zinc and copper. Histidine is therefore used as a chelating agent in some cases of arthritis, tissue overload of copper, iron or other heavy metals.

Professor Gerber of Downstate Medical Center, New York, uses between 1 g and 6 g daily in arthritic patients. Pfeiffer further notes that both histidine and histamine act as chelating agents (they will attach themselves to other substances, notably trace elements or metals) and that this may account for their usefulness in treating some forms of arthritis, where copper or other metal excess can thus be removed from the system. Pfeiffer maintains that histamine is a neurotransmitter of some as yet unspecified portion of the brain.

Pfeiffer and Iliev, of the Brain Bio Center, showed, by accurately measuring tissue histamine content, that they were able to identify two distinct categories of schizophrenia which, together, make up two thirds of those affected. The histapenic patient is extremely low in brain and blood histamine, and is usually over-stimulated. Whereas the histadelic patient is high in levels of histamine in the blood and brain, and is usually suicidally depressed. Methionine (see page 49) detoxifies histamine and so methionine is recommended by Pfeiffer for histadelic schizophrenics (i.e. those with high levels of histamine in their bloodstreams), along with calcium lactate, zinc and manganese.[2] Histidine (which goes on to become histamine) has also been shown to be useful therapeutically in people with allergic conditions, and since it acts to lower blood pressure and relax blood vessels (by acting on aspects of the nervous system) it has been used to treat a variety of circulatory and cardiac problems. Some researchers have found it a useful addition when treating anaemia because of specific beneficial effects on the blood make-up.

Histidine is known to be vital for the maintenance of myelin sheaths (the insulation which surrounds the nerves)[4] and Borrmann tells us that it is therefore useful in treating some forms of hearing disability because of its benefit to the auditory nerve. Certainly it is known that where there is a deficiency of histidine, nerve deafness is likely.[5] Pearson and Shaw point out that the release of histamines from body stores is a necessary prerequisite for sexual arousal, and histidine supplementation may assist in problems relating to this (together with niacin, and vitamin

B_6 which is required for the alterations of histidine to histamine.)[3]

Brekhman reports that as part of the Soviet space programme over 25,000 different chemical substances and compounds have been examined to try to discover effective protective substances against the effects of radiation. Among the standard preparations which are now issued to cosmonauts in this regard as nonspecific pharmacologically protective medicines is histidine (the only other amino acid is tryptophan). Dosages are not stated.[4]

Childhood requirements (RDA) are put at 33 mg per kilogram of body weight per day. It is found in animal sources of protein at levels of 17 mg per gram.[5]

Note: Since histadelic patients are displaying symptoms resulting from excessive histamine in the system, it is unwise for anyone with symptoms of manic depression to supplement with histidine unless it is established that levels of histamine are within the normal range.

References

1. Bland, J., (editor), *Medical Applications of Clinical Nutrition*, Keats, 1983.
2. Pfeiffer, Carl, *Mental and Elemental Nutrients*, Keats, 1975.
3. Pearson, D. and Shaw, S., *Life Extension*, Warner Books, 1983.
4. Brekhman, I.I., *Man and Biologically Active Substances*, Pergamon Press, 1980.
5. *Nutrition Almanac*, McGraw Hill, 1979.

Isoleucine

Although the EAA isoleucine has, as yet, not been identified as having particular therapeutic characteristics, Borrmann reports that: 'it is useful in haemoglobin formation,'[1] but he does not elaborate on that remark.

Isoleucine has been identified as one of a group of amino acids deficient in amino acid profiles run on mentally and physically ill patients,[2] as reported by Jon Pangborn Ph.D. and William Philpott, M.D. Therapeutic doses of between 240 mg and 360 mg daily are suggested in combination with the other amino acids found lacking (e.g. valine, leucine, tyrosine, cystine, glutamic acid and ketoglutaric acid). As mentioned in the previous chapter isoleucine is, as one of the branched-chain amino acids, one of the culprits in the acidemias, such as Maple Syrup Urine Disease.

Bland[3] gives the range of isoleucine requirement in normal adults between 250 mg and 700 mg daily, as against the National Academy of Sciences RDA for an adult of 12 mg per kilogram of body weight, which for a 75 kg man would mean a daily intake of around 900 mg. The isoleucine content of protein, of animal origin, is 42 mg per gram of protein.[4]

Major food sources of isoleucine are beef, chicken, fish, soy protein, soyabeans, eggs, liver, cottage cheese, baked beans, milk, rye, almonds, cashews, pumpkin seeds, sesame seeds, sunflower seeds, chickpeas, lentils.[4]

References

1. Bormann, W., *Comprehensive Answers to Nutrition*, New Horizons, Chicago, 1979.
2. Philpott Medical Center, pamphlet 'Selective Amino Acid Deficiencies'.
3. Bland, J., (editor), *Medical Applications of Clinical Nutrition*, Keats, 1983.
4. *Nutrition Almanac*, McGraw Hill, 1979.

Leucine

Leucine is an EAA with no particular identified therapeutic role, apart from its complicity in conditions relating to disorders of branched-chain amino acid metabolism, such as Maple Syrup Urine Disease. As with isoleucine it was found to be relatively deficient in assessments of amino acid status of groups of mentally and physically ill subjects[1] and is supplemented, together with the other appropriate amino acids (isoleucine, valine, tyrosine, cystine, glutamic acid and ketoglutaric acid) at a dosage of between 240 mg and 360 mg daily, in divided dosage.

The range of human requirements in health is given[2] as from 170 mg to 1100 mg daily representing a 6.4 fold possible variation in need. This was derived from a sample of only 31 individuals and so the chances of far greater variations in need existing in the public at large is great. RDA is given as 16 mg per kilogram of body weight in adults, which for a 75 kilo individual would require a daily consumption of 1200 mg. The level of leucine found in animal protein is given[3] as 70 mg per gram. Major sources of leucine in food are beef, chicken, soya protein, soya beans, fish, cottage cheese, eggs, baked beans, liver, whole wheat, brown rice, almonds, brazil nuts, cashew nuts, pumpkin seeds, lima beans, chick peas (garbanzos), lentils, corn.

Of particular interest are the reports in *The British Journal of Nutrition*[4,5] which indicate that dietary excess of leucine may be a precipitating factor in the causation of pellagra. It was found that when rats were fed on a diet that provided 15 g of leucine per kilogram in excess of requirements for a period of seven weeks it led to a significant reduction in concentrations of vitamin B_3 metabolites in the blood and liver. This effect was only apparent when the overall diet provided less than an adequate amount of nicotinamide (B_3), so that the animals were dependent upon synthesis of nicotinamide from tryptophan to meet all or part of their needs. It was noted that other nutrients and their enzymes were not thus affected by the loading of leucine to the diet. The second report, which confirms the essentials of the first, established the minutiae of the process.

Too much leucine in the diet seems to interfere with numerous biochemical processes, leading to a marked reduction in the final forms of vitamin B_3 and hence to pellagra (see discussion of Hartnup disease page 34). We can deduce that leucine *is* an essential amino acid, in that we have to derive at least some of it from our diet, but that too much of it produces imbalances which can have serious consequences. Corn has a very high level of leucine in relation to its isoleucine content, and has therefore been investigated as a factor in causing pellagra, in some countries where it is a staple food.[6] Leucine would seem to be essential, but with potentials for causing harm if other factors permit. The ratio of amino acids to each other is patently important as is the necessity for ensuring overall nutrient status, as evidenced by the fact that leucine excess had no harmful effect if nicotinamide was adequately present, whereas it was able to induce pellagra-like symptoms when the body was obliged to utilize precious tryptophan supplies to manufacture vitamin B_3. The right handed d-form of leucine has been shown to have a similar effect to that displayed by d-phenylalanine (see page 52) in that it retards the breakdown of the natural pain killers of the body, the endorphins and enkephalins. It has, however, not been researched in this regard, as has d-phenylalanine, but may in time be found to be just as useful in chronic pain control.[6]

References

1. Philpott Medical Center, pamphlet, 'Selective Amino Acid Deficiencies'.
2. Bland, J., (editor), *Medical Applications of Clinical Nutrition*, Keats, 1983.
3. *Nutrition Almanac*, McGraw Hill, 1979.
4. *B.J. of Nutrition*, Vol.49, No.3, May 1983, p231.
5. *B.J. of Nutrition*, Vol.50, No.1, July 1983, p25.
6. Pearson, D. and Shaw, S., *Life Extension*, Nutri Books, 1984.

Lysine

Lysine is an EAA and it has been found to have therapeutic effects in the viral related diseases. Particularly of current interest is its ability to control herpes simplex virus, if the diet is low in arginine. It has also been found to have other therapeutic effects which will be discussed. Lysine cannot be synthesized in the body and the breakdown of lysine is not reversible. It is therefore vital that it is in the diet in adequate quantities. Its deficiency in cereal proteins makes it the limiting factor in rice, wheat, oats, millet and sesame seeds. Insufficient intake leads to poor appetite, decrease in body weight, anaemia, enzyme disorders, etc. It is used therapeutically to enhance the growth of children, and to assist gastric function and appetite.

Lysine and Herpes

Some years ago, before herpes became such a prevalent and talked about condition, a chance observation by Dr Chris Kagan at the Cedars of Lebanon Hospital, Los Angeles, opened the way for its control by means of amino acid therapy. He noted that solutions of herpes virus cultures were always encouraged to grow rapidly by the addition of l-argenine to the solution. This was based on the research of Dr R. Tankersley, who also found that to slow down growth in a solution containing herpes virus it was necessary to add l-lysine. On the basis of this the therapeutic application of lysine was attempted. The results were excellent.[1] It was found in one trial that 43 of the 45 patients involved improved markedly. Dosages were from 300 mg to 1200 mg of lysine daily, at the same time as reducing dietary arginine intake. Patients studied for up to three years on this programme showed complete remission and no side effects. Pain disappeared rapidly, and in all cases no new vesicles appeared. Resolution of existing vesicles was more rapid than in patients' past experience. There was no extension of the initiating lesion in any cases. It was found that within one to four weeks of terminating the use of lysine, lesions returned.

Subsequent experience with this pattern of treatment has shown that providing the balance of lysine to arginine can be kept at the right levels the replication of viral particles can be checked. The failure of the method is almost always the result of inadequate lysine intake, or an excessive arginine intake, and the individual must find the correct balance by trial and error. The mechanism that is thought to operate in this control of viral particles is one in which the structurally similar lysine is absorbed into the virus instead of arginine, which is catabolized by the body. Arginine and lysine compete for transport through the intestinal wall, and if there is sufficient excess of lysine then it is successful in reducing the intake of arginine, which is required by the virus for replication.[2]

The major food sources containing a high lysine:arginine ratio are fish, chicken, beef, lamb, milk, cheese, beans, brewer's yeast and mung bean sprouts. Foods which contain a high arginine:lysine ratio, and therefore should be avoided, include gelatin, chocolate, carob, coconut, oats, wholewheat and white flour, peanuts, soybeans and wheatgerm.

Most fruits and vegetables have a lysine excess over arginine apart from peas. Vitamin C has a protective effect on body levels of lysine.[3] Dosages recommended are 500 mg to 1500 mg of lysine daily, spread through the day. Variability will depend upon the overall nutrient balance of these two substances, and biochemical individuality. During acute herpes episodes a minimum of 1500 mg of lysine, plus at least 1 gram of vitamin C (with bioflavinoids) should be taken through the day, with special attention to the dietary intake of arginine being kept low.

Lysine therapy is recommended by the Herpes Organisation in the UK, and it is available freely through health food stores and some pharmacists. Other aspects of lysine's applicability to therapeutics include the fact that from it the body forms an amino acid called carnitine which is causing some interest in its role as an agent for transporting fatty acids into body cells, where they can be used as a source of fuel in the generation of energy. If carnitine levels are low within the cells, then there is poor metabolism of fatty acids, thus contributing to an elevation of blood fat and triglycerides. Recent research[4] suggests that there is a rapid conversion, in the body, of orally-administered lysine to carnitine in humans. This may be impaired in cases of malnutrition. This will be considered further in the section dealing with carnitine (page 71).

Drs Cheraskin and Ringsdorf[5] report that deficiency of lysine results in reduced ability to concentrate. Borrmann[6] reports that it is required for antibody formation, and that deficiency results in chronic tiredness, fatigue, nausea, dizziness and anaemia. The range of human needs[7] is given as between 400 mg and 2800 mg daily, a seven-fold variation in a sample of 55 people. Adult requirements are given as[8] 12 mg per kilogram of bodyweight, which would result in a 175 lb individual requiring just in excess of 2,000 mg per day. Its availability in first class protein is approximately 50 mg per gram.

References

1. Griffith, R., Delong, D., and Kagan, C., *Dermatologica* No.156, pp257–67, 1978.
2. Yacenda, J., *The Herpes Diet*, pamphlet, Felmore Ltd., Tunbridge Wells.

3. Kagan, C. 'Lysine therapy for herpes simplex'. *The Lancet* 1:37, 1974.
4. *Am. J. Clin Ntr.* 37:Jan 1983, pp98–8.
5. *Psychodietetics*, Bantam Books, pp22, 1977.
6. *Comprehensive Answers to Nutrition*, New Horizons, Chicago, p10, 1979.
7. Bland, J., (editor), *Medical Applications of Clinical Nutrition*, Keats, 1983.
8. *Nutrition Almanac*, McGraw Hill, pp236, 1979.

Methionine

Methionine is an essential amino acid which contains sulphur and which has the ability to 'donate' part of its structure to other molecules, thus altering them. It is therefore called a 'methyl donor'. The methyl molecule (a carbon-hydrogen combination designated CH_3) which is 'donated' is essential for formation of nucleic acid (RNA/DNA) the genetic material of every cell of the body which determines the formation of all the different proteins which make up our body.

Methionine gives rise to the amino acids cysteine and cystine (see page 77). Methionine, as well as cysteine and cystine can act as a powerful detoxification agent, being capable of the removal from the body of toxic levels of heavy metals such as lead.[1] Schauss discusses the use, in such cases, of foods rich in these compounds such as beans, eggs, onions and garlic, but states: 'Since it requires large quantities of these foods to have a significant impact upon the body's toxic metal burden, it is often more desirable to use specific nutritional supplements.'

The sulphur amino acids are also noted as protectors against the effects of radiation.

Methionine is an antioxidant, and as such is a good free radical scavenger.[2] Because it has a methyl group to offer it can combine with active free radicals which are harmful to the system. Studies[3] show that alcohol is one oxidant which can stimulate the release of dangerous superoxide radicals. Methionine has shown protective effects against alcohol in this regard and in general. Methionine also aids in the maintenance of the pool of glutathione peroxidase, the powerful enzyme antioxidant.

Adelle Davis considered methionine to be 'one of the body's most powerful detoxifying agents'.[4] Pfeiffer also notes[5] its ability to detoxify histamine when levels of this are high in schizophrenic patients (histadelic).

Deficiency of methionine can be the cause of choline deficiency, according to Adelle Davis, as it can retention of fat in the liver.[5]

Williams confirms this,[6] saying: 'There are certain nutrients, sometimes called lipotropic agents, which are peculiarly effective in promoting the bodily production of lecithin. Three substances of this group are methionine, choline and inositol.' He describes an experiment at Harvard in which it was found that monkeys were afflicted with atherosclerosis as a result of consuming a completely satisfactory diet, with the single exception of methionine deficiency. The blood proteins albumen and globulin, which are connected with antibodies, cannot be synthesized without the adequate presence of methionine.[7] It is thought that choline and folic acid assist methionine in its detoxification activities.

From a therapeutic viewpoint the ability of methionine to eliminate toxic metal loads would appear to be one of its prime uses. As an essential aspect of the body's ability to use selenium, methionine has also shown great importance. It is essential for the absorption, transportation and bioavailability of selenium. In humans seleno-methionine is more readily incorporated into the tissues than other forms of selenium, such as Se-selenite.[8] The range of human needs of methionine is given as between 800 mg and 3000 mg per day. This represents a 3.7 fold variation in need, based on a sample of 29 individuals.

Daily requirements are given as 10 mg a day per kilogram of body weight for all the sulphur amino acids, including methionine.[10]

It is found primarily in the following foods: beef, chicken, fish, pork, soybeans, egg, cottage cheese, liver, sardines, yogurt, pumpkin seeds, sesame seeds, lentils. Dosage, therapeutically, varies from 200 mg to 1000 mg daily. Note that methionine has a particularly distinctive odour. It is a meaty, sulphurous smell which most people find unpleasant. Methionine metabolism disorders may be indicated in the urine by the accompaniment of excess methionine, with homocystine (which is normally not detected in urine). This could be due to deficient folic acid; or defective folic acid metabolism; or deficient intestinal absorption; or impaired vitamin B_{12} metabolism. Both B_{12} and folic acid are required for enzymatic remethylation of homocystine.[11] A more likely cause of homocystineuria would be an enzyme defect, involving cystathionine B-synthase. Symptoms could include cardiovascular, skeletal and joint changes, ocular and neurological problems, as well as brittle hair, thin skin, fatty changes in the liver and myopathy. Pyridoxal phosphate (vitamin B_6) may be deficient at the same time.

A low-methionine, cysteine/cystine supplemented diet, would be indicated if there was no response to vitamin B_6 supplementation in such a case, and B_{12}, folic acid and magnesium are often indicated to normalize the methionine metabolism in such conditions.

It is vital that the relationship between protein in general, and methionine specifically, with vitamin B_6 (pyridoxine) be understood. Methionine is an anti-oxidant. However its derivative homocysteine is a powerful oxidant. Adequate levels of vitamin B_6 allow this to be reconverted into an antioxidant substance, cystathione. A high meat intake, for example, with an inadequate vitamin B_6 intake would produce just such a situation, as would high methionine supplementation without B_6 supplementation. Cardiovascular disease could well result from such an imbalance of nutrients and consequent free radical activity in the absence of antioxidants.[12]

References

1. Schauss, Alexander, *Diet, Crime and Delinquency*, Parker House, Berkeley, 1981.
2. Passwater, R., *Supernutrition*, Pocket Book, New York, 1976.
3. *Amino Acids*, pamphlet Dietary Sales Corporation, Indiana.
4. Davis, Adelle, *Let's Eat Right to Keep Fit*, George Allen and Unwin, London, 1961.
5. Pfeiffer, Carl, *Mental and Elemental Nutrients*, Keats, 1975.
6. Williams, Roger, *Nutrition against Disease*, Bantam Books, New York, 1981.

7. Borrmann, W., *Comprehensive Answers to Nutrition*, New Horizons, Chicago, 1979.
8. *Reviews in Clinical Nutrition*, Vol.53, No.1, Jan 1983.
9. Bland, J., (editor), *Medical Applications of Clinical Nutrition*, Keats, 1983.
10. *Nutrition Almanac*, McGraw Hill, 1979.
11. Stanbury et al, *Metabolic Basis of Inherited Diseases*, McGraw-Hill, 1983.
12. Pearson, D. and Shaw, S., *Life Extension*, Warner Books, 1984.

Phenylalanine

Phenylalanine is an EAA which has been found to have remarkable therapeutic properties, and which itself gives rise to other amino acids which are the forerunners of many vital substances in the economy of the body.

Deficiency of phenylalanine can lead to a variety of symptoms, including bloodshot eyes, cataracts[1] and, according to Hoffer, many behavioural changes.[2] Hoffer points out that a number of neurotransmitters derive from phenylalanine. It is converted into tyrosine (see page 73) unless the patient is suffering from phenylketonuria (see Chapter 5). This condition occurs when the enzyme which converts phenylalanine to tyrosine is deficient. Children thus affected display psychotic behaviour, and adults typically schizophrenic behaviour.

Tyrosine is converted into norepinephrine and subsequently epinephrine (previously called noradrenalin and adrenalin). All the end-products of phenylalanine are themselves converted into other end-products, one of which is adenochrome which is a powerful hallucinogen. Adenochrome was the basis for Hoffer and Osmond's hypothesis of schizophrenia, which led to the use of niacin and vitamin C in its treatment.[3] It can be seen, therefore, that deficiency of phenylalanine can lead to a wide variety of behavioural changes. The direct conversion of phenylalanine to tyrosine, and then to dopamine and on to norepinephrine and epinephrine, indicates the wide range of potential influence that it has. Neither phenylalanine nor tyrosine should therefore be supplemented in individuals taking mono-amine-oxidase drugs (MAO's).

One of its other roles has been shown to be its involvement in the control of appetite. It has been demonstrated[4] that free amino acids in the gut, especially tryptophan and phenylalanine, trigger the release of cholycystokinin (CCK). It has been found that in man a single high protein meal, or high carbohydrate meal, can increase CCK levels from 700 pg to 1100 pg/ml within half an hour. It is thought that CCK may induce satiety (a feeling of having eaten enough), and a termination of eating, either by altering gastro-intestinal function (e.g. gastric emptying) or by interaction with central nervous system feeding centres. It is known that CCK effects of satiety do depend upon intact vagal nerve fibres, and it is thought that this route might allow CCK to interact with the brain, via CCK receptors on vagal nerve fibres. Phenylalanine is being employed as an appetite suppressant in obesity. It is taken prior to a meal to initiate CCK release and among its other effects are, frequently, a feeling of greater alertness, increased sexual interest, memory enhancement and, after 24 to 48 hours, an antidepressant effect. Pearson and Shaw[12] suggest that in the use of phenylalanine for weight reduction purposes, between 100 mg and 500 mg should be taken in the evening on an empty stomach just before retiring. This should only be continued until weight reduction is satisfactorily achieved.

It should be noted that if overall amino acid intake is low (e.g. low protein diet) and phenylalanine is taken in large doses, thus causing amino acid imbalance, there could be an induced tyrosine toxicity. In animal trials it was found that a low protein diet, combined with a level of phenylalanine equal to 3 per cent of the diet, resulted in signs of depression and eye lesions. This level of phenylalanine consumption would be difficult to achieve in man, but the possibility of incorrect use exists.[5]

Recent research reports have shown a new and potentially dramatic use for phenylalanine in the field of pain control.[6,7]

The form of phenylalanine found in the animal protein diet of man is laevo, or left-handed, phenylalanine. That found in plant and bacterial cultures is dextro, or right-handed phenylalanine. This form is converted in the body to l-phenylalanine. There also exists a so called racemic mixture consisting of equal parts of the d- and l- forms, which is known as dl-phenylalanine, or more simply DLPA. The original study reporting the pain controlling aspect of DLPA was published in 1978, by Dr Seymour Ehrenpreis and colleagues of the University of Chicago Medical School. At this stage it was d-phenylalanine that was creating interest. Patients were selected on the basis that other forms of treatment had failed. Pain relief in a variety of conditions, ranging from whiplash injury to osteo- and rheumatoid arthritis, was rapid and lasting. There were no adverse effects noted, nor was there any degree of tolerance, i.e. the pain relief did not diminish with subsequent use. Pain relief took from one week to four weeks to reach its optimum level, and frequently lasted for up to a month after the cessation of treatment.

Subsequent work by these, and other, researchers, has led to the combining of the d- and l- forms, into DLPA which not only provides the pain relieving effect but also supplies the body with its requirements of phenylalanine. The effects on arthritic conditions are especially pronounced, since the majority of cases employing DLPA have found relief. It appears from research that DLPA inhibits enzymes that are responsible for the break-down of endorphins, which appears to allow the pain relieving attributes of endorphins a longer time span for their pain relieving action. This means of course that DLPA (or d-phenylalanine on its own) is not acting as an analgesic, but is rather allowing the endogenous pain control mechanism of the body to act in a more advantageous manner.

It has been noted that patients with chronic pain problems have reduced levels of endorphin activity in the cerebrospinal fluid and serum, and DLPA (or d-phenylalanine on its own) enhances the restoration of normal levels. It is worth recording that DLPA does not interfere with the transmission of normal pain messages, thus the defence mechanism of the body is not compromised. It is only the ongoing, pain-relieving mechanism that is enhanced. DLPA is usually presented in 375 mg tablets. The usual dosage is two tablets taken 15 to 30 minutes prior to meals, to a total of six tablets daily. If there has been no improvement within a period of three weeks the dosage is doubled. If there is still no response then the DLPA should be discontinued. However, there is only a small percentage of failure (between 5 and 15 per cent). Relief is usually noted within seven days, at which time dosage is reduced in stages, until, by trial and error, the minimum maintenance dose is reached. Many patients find that a week per month on DLPA provides maintenance of pain relief; whereas others require continued taking in reduced quantities. There are apparently no contra-indications or side-effects reported to date. Since both d- and l-phenylalanine are normal constituents of the economy of the body, there is no reason why there should be any side effects, as long as the overall nutritional status is maintained.

With no contra-indications or side-effects, and with no tolerance or addiction apparent, as well as the ability for DLPA to combine with any other form of treatment, the use of this substance seems to be comprehensively assured. An antidepressant effect is also reported, which should make its use even more attractive.

The normal ranges of requirement of phenylalanine in humans is given as between 420 mg and 1,100 mg per day. This was in a sample of 38 individuals.[8]

The National Academy of Science requirement is shown as 16 mg per kilogram in adults.[9] This represents some 1,200 mg per day in a 75 kg man. Rose[10] states the recommended daily requirement to be in the region of 2.2 g. The relative difference in the figures given, indicates to some extent just how individualized dosages should be.

Food sources of phenylalanine include soybeans, cottage cheese, fish, meat, poultry, almonds, brazil nuts, pecans, pumpkins, and sesame seeds, lima beans, chickpeas (garbanzos) and lentils. The content of phenylalanine and other aromatic amino acids in first class protein is given as 73 mg per gram.[9]

Therapeutic Dosages: A general consensus suggests that depressive states are relieved within a few days by the taking of 100 mg to 500 mg of l-phenylalanine per day. Caution should be employed in the use of phenylalanine in hypertensive individuals, and low doses (around 100 mg daily) should be used at the start of a programme by anyone with suspected high blood-pressure, and a check should be kept on pressure levels.

References

1. Davis, Adelle, *Let's Eat Right to Keep Fit*, George Allen and Unwin, London 1961.
2. Bland, J., (editor), *Medical Applications of Clinical Nutrition*, Keats, 1983.
3. Hoffer, A. and Osmond, H., *How to Live with Schizophrenia*, Johnson, London, 1966.
4. *Reviews in Clinical Nutrition*, Vol.53, No.3, pp169, March 1983.
5. *Agric. Biology and Chemistry*, Vol.46, No.10, pp2491, 1982.
6. Bonica et al, *Advances in Pain Research and Therapy*, Vol.5, Raven Press, N.Y., 1983.
7. *Proceedings of International Narcotic Research Club Convention*, Ed. E. Leong Way, 1979.
8. Bland, J., (editor), *Medical Applications of Clinical Nutrition*, Keats, 1983.
9. *Nutrition Almanac*, McGraw Hill, 1979.
10. Rose, W., 'Amino Acid Requirements in Man', *Nutrition Reviews*, Vol.34, No.10, 1967.
11. *American J. of Psychiatry* 147:622, May 1980.
12. Pearson, D., and Shaw, S., *Life Extension*, Warner Books, 1984.

Threonine

Threonine is an EAA. As yet few therapeutic roles are evident. Threonine (along with lysine) is deficient in most grains and it requires the combining of a pulse which contains threonine (and lysine) with a grain, to ensure a complete protein in vegetarian meals.[1]

Deficiency in threonine results in irritability and generally difficult personality, according to Cheraskin.[2] Williams lists it[3] along with most of the B vitamins, magnesium, ascorbic acid, iodine, potassium, tryptophan, lysine and inositol and glutamic acid as being essential in mental illness prevention and treatment. Borrmann[4] states that threonine is 'very useful in indigestion and intestinal malfunctions, and prevents excessive liver fat. Nutrients are more readily absorbed when threonine is present.' Threonine serves as a carrier for phosphate in the phosphoproteins. A fatty liver, resulting from a low protein diet, will be corrected by threonine which acts to break down fat.

breaks down fat.

Research on mice indicates that variations in individual amino acid quantities in the diet can reduce the suscepti-bility of the animal to particular infections. Weaning mice fed on diets which were 75 per cent limited in histidine, or threonine, but not in methionine, were more susceptible to infection by salmonella typhimurium, whereas mice on a diet 75 per cent limited in methionine and threonine were more susceptible to infection by listeria monocytogenes. Replenishment with the limiting amino acid, histidine and threonine, reversed the susceptibility to S. typhimurium. The extrapolation of these types of nutrient imbalance to the human model could indicate ways of minimizing risks for susceptible individuals to specific infections.[7]

The range of human requirements is stated to be between 103 mg and 500 mg daily. This represents a range of 4.8 fold difference in a sample of 50 people.[5] Daily requirement is stated to be 8 mg per kilogram of body weight in adults. Its availability in first class protein is 35 mg per gram. This level of requirement would mean that a 75 kilo individual would require 600 mg per day.[6]

References

1. Davis, Adelle, *Let's Eat Right to Keep Fit*, George Allen and Unwin, London, 1961.
2. Cheraskin and Ringsdorf, *Psychodietetics*, Bantam, 1976.
3. Williams, R., *Nutrition Against Disease*, Bantam, 1981.
4. Borrmann, W., *Comprehensive Answers to Nutrition*, New Horizon, Chicago, 1979.
5. Bland, J., (editor), *Medical Applications of Clinical Nutrition*, Keats, 1983.
6. *Nutrition Almanac*, McGraw Hill, 1979.
7. *Nutrition Research*, Vol.12, No.3, pp309–17, 1982.

Tryptophan

NOTE Tryptophan in any strength is currently withdrawn from sale. The reasons for this are explained in Chapter 8.

Tryptophan is an EAA. Among its many therapeutically significant roles its essential part in the synthesis of nicotinic acid. In its own right it has been used therapeutically in the treatment of insomnia, depression and obesity.

There are, however, cautionary signals coming from a number of research centres, pointing to the necessity of tryptophan being used with care. For many years now, in its indicated areas of use (see below), and in proper relationship with other nutrients, it has been used perfectly safely. It is however capable of causing marked side effects when incorrectly used (or when contaminated, see Chapter 8).

Tryptophan is a nutrient affecting neurotransmitter function after it is converted into serotonin, which is a neurotransmitter. This can stimulate nerve cells, which amplify the transmission of signals to the cell which they are innervating.[1] A great deal of research has been conducted into the mechanisms whereby brain function is altered in relation to blood levels of the nutrient factors which influence neurotransmitter production. These include tyrosine, which becomes ultimately epinephrine; and lecithin in its pure form of phosphatidylcholine, which becomes the neurotransmitter, choline; as well as tryptophan. It has been found that serotonin levels in the blood influence the individual's choice of food, so that more or less carbohydrate will be consumed. Wurtman,[2] who has researched this area exhaustively, has found that by altering levels of carbohydrate eaten it is possible to increase the levels of serotonin in the brain. Tryptophan levels in the brain, ready for conversion to serotonin, depend upon blood tryptophan levels as well as the ratio between this and tyrosine, phenylalanine, leucine, isoleucine and valine (large neutral amino acids). Since a high protein meal leaves much less tryptophan free to be absorbed into the brain than other amino acids, less tryptophan is carried across the barrier. A high carbohydrate meal which causes insulin release has a marked effect on the five amino acids mentioned because they are circulating as free molecules. Tryptophan is, however, not in this form and is therefore unaffected by the insulin. The ramifications of this effect of food choice on the levels of serotonin have implications for the control of excessive eating.

Animals given a choice between carbohydrate or protein-rich meals not only regulate the amount of calories consumed, but also seem to control the ratio between protein and carbohydrate. Administration of a small carbohydrate-rich meal increases the level of serotonin in the brain, and this in turn increases the amount of protein in relation to carbohydrate eaten at the subsequent meal. If tryptophan is given before a meal a similar result may be anticipated since serotonin levels will rise and reduced calorie intake, via a higher protein, lower carbohydrate meal, will result, voluntarily. The phenomenon of carbohydrate craving, found in many people on a reducing diet based on a high protein diet, may therefore be the result of reduced serotonin, due to the high protein intake.

The symptoms of anxiety, tension or depression, mentioned by many people prior to a carbohydrate snack, and the relief felt afterwards, may be the direct result of relative serotonin lack followed by serotonin being thus released into brain circulation.[3,4]

A high protein meal has the opposite effect since plasma levels of the Large Neutral Amino Acids increase proportionately more than tryptophan, thus reducing the amount of free tryptophan available for crossing the blood-brain barrier, and ultimate serotonin production.

The use of this knowledge in constructing nutritional patterns which will encourage self determined weight loss is most important. To recapitulate: by giving a small quantity of carbohydrate prior to the meal it was shown that overall carbohydrate intake decreased voluntarily. If this is accompanied by, or replaced by, the intake of tryptophan then serotonin production is more assured, enhancing the likelihood of a lower carbohydrate, higher protein selection being made subsequently.[5]

Tryptophan's role in certain mental disorders involves its complex relationship with other nutrient factors. The enzyme nicotinamide-adenine dinucleotide (NAD) is required in the brain to perform several vital functions. In schizophrenics there seems to be inadequate NAD in the brain. NAD is formed by the action of Vitamin B_3 (niacin) on tryptophan, and if niacin is deficient, this transformation of tryptophan is prevented, leading to not only inadequate amounts of NAD, but to excessive amounts of tryptophan in the brain. This can lead to perception and mood changes. Pyridoxine (B_6) is also involved in the tryptophan-niacin interaction. These problems are relatively easily corrected by the supplementation of niacin and pyridoxine (vitamins B_3 and B_6).[6]

Enhancement of tryptophan uptake by the brain is reported by the use of vitamin C and pyridoxine at the same time as its oral administration.[7]

Cheraskin and Ringsdorf report[8] that there is an inverse relationship between tryptophan consumption and emotional complaints. Increasing the tryptophan intake decreases the number and severity of such complaints. Research carried out showed that of a group of 66 individuals, assessed after several months of tryptophan supplementation, those who had increased from 1,001 mg per day to an average of 1,331 mg per day showed a remarkable decrease in the number of psychological complaints, whereas those who had not altered their tryptophan intake showed no change. As mentioned previously, niacin is converted from tryptophan, under the influence of pyridoxine. The ratio of conversion is one gram of niacin from 60 g of tryptophan. Tryptophan is not richly supplied in the diet, and the amount of tryptophan that can be converted to niacin is therefore not predictable, and certainly does not meet the body's daily requirement. Tryptophan that is not thus converted to niacin or serotonin remains largely bound to albumen in the blood.

Serotonin has been widely promoted as a sleep inducing agent. Its precursor tryptophan was researched in this regard by Dr E. Hartmann of Boston State Hospital. He reported,[9] 'In our studies we found that a dose of one gram of tryptophan will cut down the time it takes to fall asleep from twenty to ten minutes. Its great advantage is that not only do you get to sleep sooner, but you do so without distortions in sleep patterns that are produced by most sleeping pills.' Goldberg and Kaufman state that they replicated Hartmann's results and found that tryptophan did not in any way depress the central nervous system but 'simply allowed the body to do what it does normally under ideal conditions.'[9]

A summary of its effects on sleep was given in a study in California.[10] Firstly it was found that tryptophan was an effective hypnotic when administered at any time of the day. Further it was found that it significantly reduced the time of sleep onset without affecting the various stages of sleep. Finally it was shown that tryptophan produces a more relaxed waking state 45 minutes after ingestion and that at this stage sleep may be induced more easily if required. By combining Vitamin B_6 and magnesium with tryptophan there is an enhancement of all the effects described above.

There have been inconsistent reports as to the efficacy of tryptophan in the treatment of endogenous depression. Broadhurst reported a 65 per cent improvement in thirty-two depressive patients, after four weeks of supplementation of 4 g of tryptophan daily.[11]

McSweeney reported that a daily intake of 3 g of tryptophan together with 1 g of nicotinamide was superior to unilateral ECT (electro-convulsive therapy) administered twice weekly when treating depression.[12] Other trials however[13] have shown less than encouraging results, with negligible antidepressant effects in unipolar depressive patients, and only partial antidepressive effects on bipolar (manic-depressive) depressive states. Among the reasons presented by researchers for the negative results of these trials is the possibility that numbers involved were too small and that the time of study was too short. It also appeared that there was a maximum level of tryptophan dosage above which efficacy diminished. Higher doses than 6 g daily could be influenced by many factors, which could affect serotonin function. The question was also raised as to the relative numbers of unipolar and bipolar depressives in the trials as they might respond differently to tryptophan. The importance of administering tryptophan well away from the consumption of protein meals was also emphasized as being a factor to stress in all future trials. It is not yet clear in what way other antidepressive agents interact with tryptophan.[14]

Buist[15] discusses the differences, which are known to exist, in two subgroups of depressive patients. These can be separated according to their therapeutic response to various antidepressants, and to their level of a norepinephrine metabolite MHPG. The first group has low urinary MHPG (and therefore low brain norepinephrine) and they do respond to amitryptaline, but do not show a favourable response to tricyclic drugs (which raise the brain norepinephrine levels, rather than serotonin). These individuals also exhibit mood elevation after taking dextroamphetamine.

The other group has a normal urinary level of MHPG (or it may be high), indicating normal or high brain levels of norepinephrine. They fail to respond to tricyclic drugs, but show a favourable response to amitryptaline which enhances brain serotonin levels, as against dopamine or norepinephrine. These individuals fail to show mood elevation in response to dextroamphetamine. It is to be expected that these two subgroups would respond differently to tyrosine (see page 73) and to tryptophan. Tyrosine raises brain levels of norepinephrine and so would be expected to improve the first group, whereas tryptophan raises serotonin levels and therefore would be expected to improve the second subgroup. By assessing MHPG levels, and previous known response to drugs such as tricyclics, it should therefore be possible to predict which depressive patients would respond to tryptophan.

The lesson to be learned from this is that whilst nutrient substances are part of the overall economy of the body, and whilst in certain conditions they can have therapeutic effects, this does not make them universally applicable in any named condition. Since quite obviously the same manifestation of dysfunction, e.g. depression, can be the result of a variety of causes and quite probably is the result of several of these, rather than just one, no single nutrient will be the means of resolving all such cases.

The correct use of any nutrient in health or ill health is to provide the body with its needs on a cellular level. If the biochemical requirement of an individual is for tryptophan, then understanding its physiological and therapeutic roles, as well as possible complications and interrelations with other nutrients, will enable its safe and successful use.

There are warnings of one hazard in the use of tryptophan

as a supplemented nutrient, and that is in the case of pregnancy. Trials on hamsters[17] have shown quite clearly that, in animals at least, a high intake of tryptophan, combined with a high protein diet, leads to reduced litter size and increased mortality. A recent report on this subject states: 'It is often assumed that a substance is safe for consumption if it occurs naturally within the body. However, such a rationale has limits with respect to tryptophan.' These trials tested tryptophan in relation to normal and high protein intake. Since a low protein diet favours transport of tryptophan to the brain and kidneys, it is to be expected that the effects of supplementation of tryptophan on pregnancy would be even more marked in such cases.

There is as yet no evidence linking tryptophan usage with any human complications of pregnancy. However, the warning is clear that until such time as it is shown to be otherwise the use of tryptophan in women anticipating becoming pregnant should be limited. Tryptophan normally occurs as 1 per cent of the protein intake, whether of plant or animal origin. In the trials quoted the levels given ranged from 3.7 per cent to 8 per cent, which it is claimed is within the range utilized in supplementation for depression and insomnia.

It should be noted that tryptophan is not compatible with monoamine oxidase inhibiting drugs.

Tryptophan can be utilized to assess the adequacy, or otherwise, of body pyridoxine levels (B_6). This is based on the concept that adequate metabolism of tryptophan requires sufficient B_6. Should there be a deficiency, or insufficiency, of B_6 then after an intake of tryptophan there would occur a urinary spill of the tryptophan metabolite xanthurenic acid. With adequate levels of B_6 an intake of between 2 g and 5 g of tryptophan produces no spillage of xanthurenic acid. In clinical studies between 50 mg and 100 mg of tryptophan per kilogram of body weight have been used to assess B_6 status. Ideally a 24 hour sample is used, but if inconvenient then the sample is collected for a six hour period following the tryptophan load. If there is an excess of 25 mg of xanthurenic acid in that six hour urine sample then B_6 insufficiency is indicated, and 75 mg of xanthurenic acid in a 24 hour sample has the same interpretation.[17] In trials, with implications for humans, but conducted on rats, previously non-aggressive rats were placed on a low tryptophan diet. They displayed aggressive tendencies after 90 days, which was unrelieved by niacin supplementation (they had also been deprived of adequate niacin). Normality was restored after sixty days either on a normal diet or intraperitoneal tryptophan injection.[17] The range of human needs of tryptophan is between 82 mg and 250 mg daily in a sample of fifty people.[18]

National Academy of Sciences give the daily requirement for an adult as 3 mg per kilogram of body weight, meaning that a 75 kg individual would need an intake of 225 mg daily. The level of tryptophan found in first class protein (and plant protein) is approximately 11 mg per gram.[19] The best sources of tryptophan from food can be found in the following: soya protein, brown rice (uncooked), cottage cheese, fish, beef, liver, lamb, peanuts, pumpkin and sesame seeds and lentils.

Note that vitamin B_6 is essential for the conversion of tryptophan, and pellagra is considered a combined deficiency of niacin, pyridoxine and tryptophan.[20]

A recent study at Finland's University of Tampere, Department of Neurology, indicates that tryptophan has potential as a pain reducing agent.

Eleven healthy volunteers were randomly assigned, in a double-blind crossover trial, to either 2 g of tryptophan daily, or a placebo. Dietary instructions were that a high carbohydrate, low fat, low protein diet, should be adhered to (to enhance tryptophan uptake in the brain). Pain was induced by a submaximal application of a tourniquet, to produce ischaemic pain, which was assessed before dietary changes; after tryptophan and after placebo. Blood samples were taken to assess tryptophan levels, and of the other amino acids which compete with it for uptake. There was a general tendency for pain to be reduced with tryptophan, and in two subjects remarkable increases in pain tolerance levels were noted.[21]

According to research into the most suitable timing for the taking of tryptophan,[22] one of its main limitations for uptake, and ultimate conversion to serotonin (or niacin), is the competition that it has with leucine, isoleucine, tyrosine, phenylalanine, valine and threonine. Thus, supplementation should be away from protein meals, and preferably together with a carbohydrate meal or snack. The resulting insulin release will ensure that competing large neutral amino acids are taken into the musculo-skeletal tissues, leaving a relatively greater amount of tryptophan in the blood. An hour prior to a protein meal is the closest tryptophan should be administered to protein. A snack may be as small as a single biscuit, or preferably a fruit or vegetable juice (e.g. carrot). Vitamin B_6 should be taken at the same time to maximize the serotonergic effect.

Recent research by Dr G. Chowinard of McGill University, Montreal, indicates that the functional usefulness of tryptophan is enhanced by supplementation of niacinamide at the same time.[23] The ratio suggested is two parts tryptophan to one part niacinamide.

Tryptophan has recently been withdrawn from both general sale and most prescription combinations in the United Kingdom and United States, due to alarm over an epidemic of side effects (eosinophilia-myalgia syndrome, with symptoms of muscle pain, fever, skin rashes, swellings and respiratory symptoms, and a few deaths) which are considered by most experts to relate to accidental contamination of tryptophan's raw material, by its Japanese manufacturer. Some scientists, however, believe that rather than (or as well as) contamination, there may be a combination of environmental and genetic factors interacting with tryptophan intake to cause this problem (see the report in *Pharmaceutical Journal*, April 21, 1990 p.486). For a deeper examination of this crisis see Chapter 8.

References

1. *Scientific American*, April 1982, pp50–58.
2. *Lancet*, 1 May 1983, pp1145. *American J. of Clinical Nutrition* Vol.34, No.10, p2045, 1982.
3. *Journal of Nutrition*, No.112, p2001, 1982.
4. *Reviews of Clinical Nutrition*, Vol.53, No.3, p169.
5. *Physiology and Behaviour*, No.29, p779, 1982.
6. Philpott and Kalita, *Brain Allergies*, Keats, 1980.
7. Passwater, R., *Super Nutrition*, Pocket Book, 1976.

8. Cheraskin and Ringsdorf, *Psychodietics*, Bantam, 1976.
9. Goldberg, P. and Kaufman, D., *Natural Sleep*, Rodale, 1978.
10. *Psychopharmaceutical Bulletin* No.17, pp81-2, 1981.
11. *Lancet* Vol.1 1392, 1970.
12. *Lancet* Vol.11 510-511, 1975.
13. *Psychopharmacologia* (Berlin) 34, pp11-20, 1974.
14. *Psychopharmacology Bulletin* 18, pp7-18, 1982.
15. *International Clinical Nutrition Review*, Vol.3, No.2, 1983.
16. *Life Sciences* 32:1193, 1983.
17. *Bolletino Soc. Italiana di Biologia Sperimenta* 58 (19) 1271, 1982.
18. Bland, J., (editor), *Medical Applications of Clinical Nutrition*, Keats, 1983.
19. *Nutrition Almanac*, McGraw Hill, 1979.
20. Pfeiffer, Carl, *Mental and Elemental Nutrients*, Keats, 1975.
21. *Acupuncture and Electro Therapeutics Research* Vol.8, No.2, pp156, 1983.
22. *International Academy of Nutrition Newsletter*, November 1983.
23. Mindell, Earl, *Tryptophan*, 1981.

Valine

Valine is an EAA but thus far little has been ascertained as to its therapeutic value.

In trials to assess the effects of pre-meal intake of amino acids, conducted in 1982, the combination utilized was phenylalanine (3 g), valine (2 g), methionine (2 g), and tryptophan (1 g). The results showed 4 g of the mixture, in the ratio given, resulted in reduced food intake in 50 per cent of the obese subjects. As described in the section on phenylalanine (page 52) this is thought to be the result of the release of choleycystokinin which induces a feeling of satiety. When combined with the tryptophan-induced presence of additional serotonin, and consequent feelings of drowsiness and calm, this is thought to result in a lesser desire for food.[1] What role valine plays in this formula is not clear.

There is a class of patients suffering from hypervali-naemias, or subacute b-aminoisobutyric aciduria, with symptoms ranging from headaches and irritability to 'crawling skin', and delusions and hallucinations. Symptoms may be aggravated by eating high-valine foods or taking a supplement which contains valine. Treatment is by a low protein diet, and the taking of supplements which excludes valine and methionine and histidine which all provoke the protein-intolerant syndrome which may be part of the complex of biochemical faults in such cases.[2] (Klaire Laboratories produce *Amino Complex 111*, which is available from York Nutritional Supplies in the U.K., and which corresponds to this formulated need.)

Borrmann describes valine as 'useful in muscle, mental and emotional upsets and in insomnia and nervousness'.[3]

The range of human needs is given as between 375 mg and 800 mg per day in a sample of 48 people, which shows a 2.1 fold variation.[4]

The daily requirements given by the U.S. National Academy of Sciences is 14 mg per kilogram of body weight per day in an adult. This indicates a daily requirement of 1,050 mg for a 75 kg individual. The content of valine in first class protein is 48 mg per gram.[5]

Main food sources of valine include soy flour, raw brown rice, cottage cheese, fish, beef, lamb, chicken, almonds, brazil nuts, cashews, peanuts, sesame seed, lentils, chick peas (garbanzos) and lima beans (raw), mushrooms, soybeans.

Valine is the amino acid that is genetically substituted for glutamic acid in the haemoglobin molecule, resulting in sickle cell anaemia.[6]

References
1. *American Journal of Nutrition*, Vol.34, No.10, p2045, 1982.
2. Pangborn, Jon, Ph.D, pamphlet. Klaire Laboratories, Carlsbad, California.
3. Borrmann, W., *Comprehensive Answers to Nutrition*, New Horizons, Chicago, 1979.
4. Bland, J., (editor), *Medical Applications of Clinical Nutrition*, Keats, 1983.
5. *Nutrition Almanac*, McGraw Hill, 1979.
6. Dixon-George, Bernard, *Beyond the Magic Bullet*, Allen & Unwin, 1978.

Proline

Proline is not an EAA but may be synthesized by the body. It is one of the main components of collagen, the connective tissue structure that binds and supports all other tissues. Pauling[1] points out that there is evidence that vitamin C is required for the processes which ultimately give collagen its characteristic properties.

The use of proline in wound healing, and in the promotion of improved collagen status, as well as in cosmetic improvement of 'ageing' tissues has been proposed by researchers in California.[2] Hydroxyproline, which the body incorporates into collagen, is readily transformed by the body from proline; it is incorporated into the structure of tendons and ligaments.[3]

Proline is one of the aromatic amino acids, such as phenylalanine and tryptophan.

Supplementation would seem to be indicated in cases of persistent soft tissue strains; hypermobile joints; soft tissue healing requirement, and in lax and 'sagging' tissues associated with age. Combined with vitamin C supplementation it is more effective. Pfeiffer states clearly that the protein collagen is neither properly formed, nor maintained, if vitamin C is lacking.[4]

References

1. Pauling, Linus, *Vitamin C – The Common Cold and Flu*, Freeman and Co, 1976.
2. Levine, Stephen, *Allergy Research Group Pamphlet* Concord, California.
3. Anthony Harris, *Your Body*, Futura, 1979.
4. Pfeiffer, Carl, *Mental and Elemental Nutrients*, Keats, 1975.

Taurine

Taurine is not an EAA. It is manufactured in the body and is also found in animal protein, but not in vegetable protein. It is a sulphur amino acid derivative.

Its synthesis in humans is from the amino acids methionine and cysteine, primarily in the liver with the assistance of vitamin B_6. Bland states[1] that vegetarians on a diet containing imbalanced protein intake, and therefore deficient in methionine or cysteine, may have difficulty manufacturing taurine.

Dietary intake is thought to be more necessary in women,

since the female hormone estradiol depresses the formation of taurine in the liver. Any additional estradiol in the form of medication would increase this inhibition. In animal studies large oral doses of taurine have been shown to stimulate production of growth hormone.

The main interest, until recently, has been in taurine's role as a neurotransmitter in which role it functions with glycine and gamma-aminobutyric acid, two neuroinhibitory transmitters. A further role played by taurine is in maintaining the correct composition of bile, and in maintaining solubility of cholesterol. Several studies have shown that bile acids are secreted, in bile, in a form in which they are conjugated (joined together) with glycine and taurine. The taurine conjugates are described as 'superior biological detergents'.[2] The data in a recent trial showed that increasing the availability of taurine through diet, probably exerts a protective effect against gall bladder disease.

Changes in platelet function are considered to be one possible cause of migraine, and taurine is apparently uniquely concentrated in the platelets (oval cells found in the blood, important in blood coagulation). This connection with migraine was established by assessing taurine levels during and after headaches. It was found that taurine levels in the platelets were significantly higher during headache periods. The trial[3] authors proposed that the metabolic platelet defect in migraine involves taurine as well as the tryptophan derivative serotonin.

Other researchers showed that taurine is found in the developing brain in concentrations up to four times that found in the adult brain.[4] Since taurine acts as a suppressor of neuronal (nerve) activity in the developing brain, during the phase when other regulatory systems are not fully developed, it is thought that deficiency of taurine, at this stage, might contribute towards, or predispose the individual to, epilepsy. Taurine has been shown in human trials to have an anticonvulsive effect.[5] Its apparent role is that it normalizes the balance of other amino acids, which in epilepsy are thoroughly disordered. In epilepsy blood levels of over half the amino acids are lowered,[6] whilst levels of taurine are high and cerebro-spinal fluid levels are low.

Blood zinc levels have been found to be low in epileptics, and since low serum zinc results in blood and urine levels of taurine rising this may be part of taurine's association with epilepsy.

Dosage is suggested at one gram a day, not more, followed by daily doses of not more than 500 mg, and reducing to 50 mg and 100 mg a day. High doses are not as effective as low doses repeated infrequently, since taurine accumulates rapidly and is only slowly metabolized. Full spectrum light exposure results in increased levels of taurine being concentrated in the pineal and pituitary glands.[7] Continued exposure to artificial lighting, which is deficient in the ultraviolet portion of the spectrum, might cause this concentration to be reversed, and to impair whatever function taurine performs in the pituitary and pineal glands. Taurine is associated with zinc in eye function, and impairment of vision has been shown with taurine deficiency prior to the development of structural changes.[8]

Taurine has also been shown to have a role in sparing the loss of potassium in heart muscle. It is thought to be the substance regulating osmotic control of calcium, as well as potassium, in heart muscle. This has been shown to be of importance during dieting periods for weight loss. During any stringent dieting programme the addition of sulphur-rich amino acids, such as methionine and cysteine, will ensure adequate taurine and therefore protect the heart muscle from calcium and potassium loss.

Taurine has been found to have an influence upon blood sugar levels, similar to that of insulin.[9] Bland comments[1] upon the ubiquitous role of taurine, which he dubs 'a remarkable accessory food factor'. He points to its possible involvement with muscular dystrophy, where its interrelationship with vitamins A and E is thought to be of importance. He discusses also the link with Down's syndrome children, in which IQ levels are said to have improved with taurine supplementation (along with B complex, C and E vitamins).[10] Should there be a genetic, or metabolic defect in the individual's ability to synthesize taurine, then supplementation could become critical. There are a number of metabolic disorders that can result in taurine levels in urine being high (apart from dietary oversupply of it, or its precursors methionine and cysteine). Impaired kidney conservation may be responsible. If other beta amino acids, such as GABA, are being normally excreted, then transport disorders can be ruled out. Heart conditions, such as myocardial infarction, or skeletal damage, physical or emotional stress, and various blood diseases, are all potential causes of increased taurine excretion in the urine. High alcohol consumption, and the use of salycilates (aspirin) may also be implicated, as can a deficiency of zinc, impairing as a result the integrity of cell membranes. Gastro-intestinal pain, acute cholecystitis (gall bladder disease) and cardiac arrythmias, may all accompany high urine levels of taurine.[11]

References

1. *International Clinical Nutrition Review*, Vol.2, No.3, 1982.
2. *Am J. of Clin. Nutrition*, Vol.37, No.2, p221, 1983.
3. *Headache Journal*, Vol.22, No.4, pp165, 1982.
4. *Orthomolecular Review*, Vol.3, No.3, 1983.
5. *Taurine*, Ed. Huxtable, Barbeau, pp1–9.
6. *Epilepsia*, 16:246–9, 1975.
7. *Life Sciences*, 22:1789–98, 1978.
8. *Nature*, 194:300–2, 1962.
9. *Canadian Chemical Process Industry*, 26:569–70, 1942.
10. *Proceedings Nat. Acad. Sciences*, 78:564–78, 1978.
11. Huxtable, R. and Pasante-Morales, H., *Taurine in Nutrition and Neurology*, Plenum Press, 1982.

Carnitine

Carnitine is synthesized in the liver by humans as well as being a part of the diet in the form of muscle and organ meats. It is not found in vegetable forms of protein. Carnitine is not an EAA.

A number of therapeutic roles have been described for carnitine which is converted rapidly from lysine as well as methionine.[1] The process of conversion is dependent upon adequate vitamin C being present.[2] The supply of carnitine is especially enhanced by lysine ingestion, as compared with other amino acid precursors of carnitine such as threonine and tryptophan.[1]

It is suggested that men have a higher need for carnitine than women. Higher levels are found in blood in men than women, and men have high levels present in the epididymis of the testes. Lysine depletion in animals results in infertility as a result of the loss of sperm motility.[3,4] Bland[2] suggests that although carnitine is not a vitamin it may be an essential nutrient in newborn infants, due to inadequate ability to synthesize it; and in adults with genetic limitations in their ability to convert methionine or lysine to carnitine.

Carnitine has been shown to have a profound involvement in the metabolism of fat, and in the reduction of triglycerides. Oxidation of triglycerides occurs when 1 g to 3 g of carnitine are administered daily. This is of potential value in conditions as diverse as intermittent claudication; poor hand and foot circulation; myocardial infarction[5] and kidney disease. Carnitine transfers fatty acids across the membranes of the mitochondria (energy producing centres in all body cells), where they can be utilized as sources of energy.

A variety of other conditions have been suggested as being potential beneficiaries of carnitine supplementation, including muscular dystrophy, myotonic dystrophy, and limb-girdle muscular dystrophy, since these lead to carnitine loss in the urine and therefore greater requirements.[6] The application of the use of carnitine to the stimulation of fat metabolism leads to possible benefits in cases of obesity. Since fat is more readily mobilized, and clearance is more rapid, with the use of carnitine, there is every reason to expect that a clinical application in this direction will be forthcoming with further research.

Research in Rome[5] showed that during acute, or chronic, cardiac ischemia, or chronic hypoxia, there occurs an accumulation of free fatty acids and long chain acyl-CoA-esters which can damage the myocardium. Carnitine appears to offer protection by forming compounds with these fatty substances. Carnitine has been shown to be deficient in hearts of patients who have died of acute myocardial infarctions, especially in damaged tissue. If carnitine were available, then, it is thought, the areas immediately surrounding the damaged areas could be restored to normal.

Carnitine has been shown[7] to be useful in conditions of ketosis (build up of acid wastes in the blood) in individuals on diets which produce the accumulation of ketone bodies, or fat waste products, in the blood. Such a build up can acidify the blood, resulting in calcium, magnesium and potassium loss, and can indeed be life-threatening. Fat metabolism requires carnitine to be adequately present. In scurvy the fat levels of the blood are high,[8] and this is thought to be as a result of the relationship which exists between vitamin C and carnitine. A low level of vitamin C will result in apparent carnitine deficiency.

References

1. *Am. J. of Clin. Nutrition*, Vol.37, No.1, p93, 1983.
2. *International Clinical Nutrition Review*, Vol.2, No.3, p14, 1982.
3. *Clinical Chem. Acta.*, 67:207–12, 1977.
4. *Journal of Nutrition*, 107:1209–15, 1977.
5. *Lancet*, Vol.1, pp1419–20, 1982.
6. *Am J. Clin. Nut.*, 34:2693–8, 1981.
7. Earl Mindell, *Carnitine*, 1981.
8. Hulse, J. et al, *Journal of Biological Chemistry*, No.253, pp1654–9, 1978.

Tyrosine

Tyrosine is not an EAA. Tyrosine is a precursor to thyroid, adrenocortical hormones and to dopamine. Some of the symptoms of its deficiency include low body temperature, low blood-pressure and 'restless legs'.[1] Tyrosine derives from phenylalanine. Tyrosine is capable of producing toxic reactions, in excessive dosages of itself, or of phenylalanine,[2] as has been demonstrated when rats fed a low protein diet, which contained more than 3 per cent phenylalanine, developed lesions on paws and eyes, and had growth and food intakes depressed, all of which is identical with tyrosine toxicity. The pigment of skin and hair, melanin, is derived from tyrosine.

Therapeutically tyrosine has been employed to enhance its derivatives (dopa, dopamine, norepinephrine, epinephrine) as well as its ability to alter brain function. Brain tyrosine levels are most conveniently raised by ingestion of pure tyrosine, with a high carbohydrate meal to lower levels of competing amino acids. A high protein meal will increase blood and brain tyrosine to a degree but not enough to effect neurotransmitter synthesis greatly. Physiologically active neurons are highly responsive to neurotransmitters such as tyrosine (and choline) and will actively synthesize neurotransmitters from these substances. There are very few side effects resulting from even fairly large doses of tyrosine.[3] Wurtman who has done much research into this area of biochemistry suggests that if neurons (nerve cells) are not active the precursor is not used. If the neurons are active however, a particular dose of tyrosine can either reduce blood pressure in hypertension, or increase it in haemmorrhagic shock, by virtue of the provision of the tyrosine for neurotransmitter synthesis, these active neurons, will then produce the particular physiologically desirable effect.

Tyrosine is reported to help some Parkinson patients, and to aid in relieving some depression cases.[3] The use in depression of such drugs as monoamine oxidase inhibitors and tricyclic antidepressants involves an increase in the brain's levels of substances such as serotonin and norepinephrine, either by slowing down their degradation, or by prolonging their action. The use of neurotransmitter precursors, which can increase the levels of serotonin and norepinephrines is another way of achieving a similar end. The use of their precursors tryptophan (see page 57) and tyrosine can therefore be seen to be a logical step in this direction. Tyrosine has been found to be most effective[4] when there exists a deficiency state. Patients who have previously responded to amphetamines may respond well to tyrosine therapy.

There is evidence that small doses of tyrosine are more effective in increasing brain levels of neurotransmitters, than large doses.[5] Although blood and brain levels of tyrosine will increase with large doses, there appears to be an inhibition of the enzyme tyrosine hydroxylase which converts tyrosine to neurotransmitters, when large amounts of tyrosine are present.

Research into tyrosine is in its early stages and more will be heard of this powerful substance in human economy.

Note: Tyrosine is not compatible with the taking of MAO drugs (monoamineoxidases).

References

1. Philpott, W., pamphlet on *Selective Amino Acid Deficiencies*, Klaire Laboratories, California.
2. *Agricultural Biology and Chemistry*, Vol.46, No.10, p2491, 1982.
3. *Lancet*, p1145, 21 May, 1983.
4. *Psychopharmacology Bulletin*, No.18, pp7–18, 1982.
5. *Biochemical Journal*, Vol.206, pp165, 1982.

Glutamine and Glutamic Acid   *brain fuel*

Glutamine is not an EAA although Bland suggests that under certain conditions it may become a 'contingency nutrient' and therefore essential.[1] He points out that glutamine is synthesized in certain tissues for use in others, and that it is the dominant amino acid in blood and cerebrospinal fluid. It is the only amino acid that easily passes the blood-brain barrier. Glutamic acid can be synthesized from a number of amino acids including ornithine and arginine. It easily loses its amine group, and thus participates in reactions which are vital in the process of the formation of NEAA's. When glutamic acid combines with ammonia it becomes glutamine. It is also a major excitatory neurotransmitter in the brain and spinal cord, and is the precursor of GABA, which is an inhibitory transmitter, as well as glutathione (see page 85).

Williams[2] comments that the glutamine derivative glutamic acid does not pass this blood-brain barrier, although it might be in the blood in relatively high levels, and yet infiltrate the brain fluids in only small amounts. Its amide, glutamine, has no such problem, after which it is readily converted into glutamic acid. The function is described by Williams thus: 'The essential and suggestive fact to remember is that glutamic acid is uniquely a brain fuel.' Pfeiffer[3] shows that vitamin B_6 allows the removal of the acid group from glutamic acid, and the formation of gamma aminobutyric acid (GABA), which is a calming agent, and possibly a neurotransmitter. It is also a key component of the chromium compound known as the glucose tolerance factor (GTF). The best source of this, for anyone suspected of glucose intolerance, is Brewer's yeast.

Behavioural problems and autism in children have been successfully assisted by Dr Bernard Rimland, of the Institute for Child Behaviour, by nutritional means, which include glutamic acid as a major component.

Glutamic acid, having been converted to that form in the brain from glutamine, is involved in two key roles. Along with glucose it is the fuel for the brain cells.[4] The second function of glutamine is to act as a detoxifier of ammonia from the brain. As it picks up ammonia, glutamic acid is reconverted to its original form of glutamine. Passwater points out[4] that as the brain is able to store relatively small quantities of glucose, it is dependent upon glutamic acid. He states: 'The shortage of l-glutamine in the diet, or glutamic acid in the brain, results in brain damage due to excess ammonia, or a brain that can never get into "high gear".' Dr William Shive has pointed out that glutamine has a protective role to play in the body's relationship with alcohol. It has been shown to protect bacteria against alcohol poisoning,[5] and when given to rats it decreases their voluntary alcohol consumption. It is the only substance to have this effect. Williams[2] assumes this to be the result of its effect, in the brain, on the appetite centre. Glutamic acid has no such protective effect on bacteria, and presumably doesn't on humans either. Williams suggests between 2 g and 4 g daily of glutamine, as a treatment for anyone with an alcohol problem. Passwater comments on a case in which glutamine has stopped sugar craving in much the way that it has been shown to stop or reduce alcohol craving. Presumably by a similar action on the appetite centre (in the hypothalamus).

Among other noted areas of usefulness are its application in depression; IQ improvement in mentally deficient children; enhanced peptic ulcer healing; benefits to epileptic children, and applications in schizophrenia and senility (by Dr Abram Hoffer).[6]

Dr H. Newbold recommends that to attain an optimum level of intake 200 mg should be taken three times daily for a week, increasing to two capsules of 200 mg each three times daily after that, to assess general well-being. The suggested pattern for alcohol problems is 1 g three times daily.[7] There are seldom glutamine deficiencies, according to Philpott,[8] but, as Bland explains, contingency status may be reached through excessive demand in relation to genetic factors which lead to too little being synthesized in the body.

Pfeiffer discusses the way in which the well-known 'Chinese Restaurant Syndrome' relates to glutamine. Glutamic acid, which is present in monosodium glutamate, combining with a pressor amine such as tyromine, which is commonly found in certain protein foods such as aged cheese, pickled herring etc. produces the headache. Much Chinese restaurant food is also high in salt leading to fluid retention which adds to the problem.[3] Bland notes that sensitivity to monosodium glutamate indicates a need for supplemental pyridoxine.[9] Doses of between 50 mg and 100 mg daily are suggested.

References

1. Bland, J., (editor), *Medical Applications of Clinical Nutrition*, Keats, 1983.
2. Williams, R., *Nutrition Against Disease*, Bantam, 1981.
3. Pfeiffer, Carl, *Mental and Elemental Nutrients*, Keats, 1975.
4. Passwater, R., *L-Glutamine The Surprising Brain Fuel*. Pamphlet.
5. *J. Biol. Chem.*, Vol.1.214, No.2, pp503, 1955.
6. *Orthomolecular Psychiatry*, Freeman and Co., San Francisco, 1973.
7. Newbold, H., *Mega Nutrient for Your Nerves*, Berkeley Books, New York, 1978.
8. Philpott, W., *General Amino Acid Deficiencies*, pamphlet, Klaire Laboratories, California.
9. Pearson, D. and Shaw, S., *Life Extension*, Warner Books, 1983.

Cysteine and Cystine

Cystine is a stable form of the sulphur-rich amino acid cysteine. The body is capable of converting one to the other as required. In metabolic terms they can be thought of as the same.

Apart from methionine, all the sulphur-rich amino acids can be synthesized by the body, from methionine and elemental sulphur. These are taurine, cysteine and cystine.

Methionine and cysteine are used in the formation of a number of essential compounds, such as coenzyme A, heparin, biotin, lipoic acid and glutathione (see page 85). Cysteine is a vital component of the glucose tolerance factor (along with glycine, glutamic acid, niacin and chromium). Cystine is found in abundance in a variety of proteins such as hair keratin, insulin, the digestive enzymes chromotrypsinogen A, and trypsinogen, papain and also lactoglobulin. The flexibility of the skin, as well as the texture is influenced by cysteine as it has the ability to slow abnormal cross-linkages in collagen, the connective tissue protein. Cysteine will convert cystine in the absence of vitamin C. There is strong caution regarding the use of cysteine by diabetics (see note at the end of this section).

The enzyme glutathione peroxidase contains a large element of cysteine. As a detoxification agent cystine has been shown to protect the body against damage induced by alcohol and cigarette smoking. One report stated that not only was it effective in preventing the side-effects of drinking, such as hangover, but that it prevented liver and brain damage as well. It also reduces damage such as emphysema, resulting from smoking.[1]

Philpott maintains that for proper utilization of vitamin B_6, cystine or cysteine is essential.[2] The measurement of cystine by 24 hour urine and blood serum studies in a variety of chronic degenerative illnesses, both mental and physical, has been correlated with B_6 utilization disorder in research by Philpott. The results show that low B_6 utilization is produced, at least in part, by low levels of cystine (cysteine).

The metabolic steps of the formation of these two amino acids is from methionine to cystathionine to cysteine to cystine. In chronic diseases it appears that the formation of cysteine from methionine is prevented. One element in the correction of the biochemistry of chronic disease is therefore the restoration of adequate levels of cysteine (or cystine). Supplementation is one method of short term correction of such a relative deficiency, and Philpott suggests a dosage of cysteine or cystine of 1 g three times daily for one month, then reducing to twice daily.

It is noted that cysteine is more soluble than cystine and

that it contributes its sulphur more readily, and thus achieves better results in some patients. The very presence of the sulphur taste, however, makes its encapsulation desirable. Philpott recommends that B_6 be taken in doses of 50 mg three times daily in the form of Pyridoxal-5-phosphate, at the same time as cysteine. These recommendations, as with all others relating to doses, must be seen in the context of biochemical individuality, and therefore subject to large variations. It should also be noted that all recommendations assume that an overall assessment of nutrient status has been undertaken. No single nutrient is seen to be curative in any condition. Levine points to cysteine and cystine being important in stabilizing crosslinks in keratine and other proteins as well as being useful for heavy metal detoxification.[3] This is common to all sulphur amino acids.

People with diabetic tendencies should not use large supplemental doses of cysteine unless under supervision, as it is capable of inactivating insulin by reducing certain disulphide bonds which determine its structure.[5] Pearson and Shaw note[5] that in order to avoid the conversion of cysteine to cystine, with possible consequences as far as the formation of kidney or bladder stones, at least three times the dose of vitamin C should accompany the taking of cysteine supplementally.

References

1. *Nutritional Consultants*, p12, Nov/Dec 1980.
2. Pfeiffer, Carl, *Mental and Elemental Nutrients*, Keats, 1975.
3. Philpott, W., *Philpott Medical Center*, Oklahoma City, pamphlet.
4. Levine, Stephen, Allergy Research Group, Concord California, pamphlet.
5. Pearson, D. and Shaw, S., *Life Extension*, Nutri Books, 1984.

Glycine

Glycine is not an EAA. It is utilized in liver detoxification compounds, such as glutathione (of which it is an essential part together with cysteine and glutamic acid). Glycine is essential for the biosynthesis of nucleic acids as well as of bile acids.[1] Glycine is a glucogenic (gives rise to sugar) amino acid.

In its own right it has not been shown to have therapeutic applications, but as a major part of a detoxification compound, such as glutathione, it is of profound importance.[2]

Glycine is a major part of the pool of amino acids which are available for the synthesis of non-essential amino acids in the body. It is readily converted into serine. It is a constituent of a number of amino acid compound formulations used as tonic preparations.

Experimental evidence on rats indicates that glycine (in conjunction with arginine) has a useful role to play in promoting healing after trauma.[3] Traumatized rats were fed on diets without, or with, glycine plus arginine, or with ornithine plus glycine. These amino acids occur in particularly high concentrations in the skin and connective tissue, and might be required for repair of damaged tissue. Arginine and glycine supplementation significantly improved nitrogen retention in both traumatized and non-traumatized animals, whereas ornithine was less effective in this role. It is thought that creatine (an important compound used in muscle contraction) synthesis, and turnover, results from the enrichment of arginine and glycine, and this produces repair beneifts.

Human trials indicate that gastric acid secretion is enhanced by glycine.[4]

References

1. Levine, Stephen, Allergy Research Group Publication, Concord, California.
2. Amino Acids pamphlet, Dietary Sales, Indiana.
3. *Journal of Nutrition*, Vol.111, No.7, pp1265–74, 1981.
4. *Am. J. of Physiology*, Vol.242, No.2, ppG85–G88, 1982.

Alanine

This is a non-essential amino acid. The main nutritional function of alanine is in the metabolism of tryptophan and pyridoxine, in which it plays an essential role. In conditions such as hypoglycaemia (low blood sugar) alanine may be used as a source for the production of glucose, in order to stabilize blood glucose, over lengthy periods.

In trials designed to assess the effect on high cholesterol levels of combinations of different amino acids, alanine was found to have a cholesterol-reducing effect in the serum of experimental animals (rats), when in combination with arginine and glycine.[1] Levels were reduced by 20 per cent when arginine and alanine alone were administered, and by a full fifty per cent when glycine was also added. Alanine is usually included in amino acid compound tablets; where daily intake levels of between 200 mg and 600 mg daily were suggested.

References

1. *Atherosclerosis*, No.43, pp381, 1982.

b-Alanine

only natural occuring B-amino acid

This is the only naturally occuring b-amino acid. It is found in its free state in the brain. It is a component of carnosine, anserine and of pantothenic acid (vitamin B_5) which is itself a component of coenzyme A. The function of carnosine and anserine (which occur in animal muscle) is unknown.

b-Alanine is metabolized to acetic acid, and in plants and micro-organisms it is formed from aspartic acid.[1]

Therapeutically it is useful to assist in synthesis of pantothenic acid (vitamin B_5).

References

1. Meister, A., *Biochemistry of the Amino Acids*, Academic Press, New York, 1965.

Gamma-Aminobutyric Acid (GABA)

This is a non-essential amino acid, formed from glutamic acid. Its function in the central nervous system appears to be as a regulator of nerve cell activity. It is essential for brain metabolism.

It has been used in the treatment of epilepsy and hypertension.[1,2] It is thought to induce calmness and tranquillity by inhibiting neurotransmitters which decrease the activity of those neurons involved in manic behaviour and acute agitation.

Pearson and Shaw[3] point out that GABA may be useful in reducing enlarged prostate problems, by virtue of the stimulation of the release of the hormone prolactin by the pituitary. Doses of 20 mg to 40 mg daily are recommended (dissolved under the tongue). This is not suggested as an alternative to seeking professional advice in problems of this sort.

References

1. *Physiology Review*, Vol.39, pp384–406, 1959.
2. *International Review of Neurobiology*, Vol.2, pp279–332, 1960.
3. Pearson, D. and Shaw, S., *Life Extension*, Warner Books, 1983.

Asparagine and Aspartic Acid

Aspartic acid is a non-essential amino acid which plays a vital role in metabolism. It is found in abundance in plant protein. It can be turned by the body into a source of sugar (energy) and is very active in the processes which turn one amino acid into another known as amination and trans-amination.

It is plentiful in plants, especially in sprouting seeds. In protein, aspartic acid exists mainly in the form of its amide, asparagine. In plants asparagine is therefore in a reversible combination of ammonia and aspartic acid. This is important in the metabolism of plants in order to preserve ammonia. Asparagine serves as an amino donor in liver transamination processes, and participates in metabolic control of the brain and nervous system. It has therapeutic uses in treatment of brain and neural conditions. Aspartic acid performs an important role in the urea cycle. Aspartic acid, as a potassium or magnesium salt, is useful in physiological cellular function. Aspartic acid is used therapeutically in the detoxification of ammonia, and to enhance liver function.[1,2]

According to researcher and author Earl Mindell, aspartic acid increases stamina and endurance in athletes. Its ability to increase resistance to fatigue is thought to be as a result of its role in clearing ammonia from the system.[3] Asparagine also plays an important role in the synthesis of glycoprotein and many other proteins.

References

1. Meister, A., *Biochemistry of Amino Acids*, Academic Press, New York, 1965.
2. Greenstein, J. and Winitz, M., *Chemistry of the Amino Acids*, Wiley, New York, 1961.
3. Mindell, E., *Three Amino Acids for Your Health*, pamphlet, 1981.

Citrulline

This exists primarily in the liver and is a major component of the urea cycle. It exists plentifully in plant foods such as onion and garlic. It is formed in the urea cycle by the addition to ornithine of carbon dioxide and ammonia. In combination with aspartic acid it forms arginosuccinic acid, which on further metabolization becomes arginine.

Therapeutically it is used for detoxification of ammonia and in the treatment of fatigue.[1] As a precursor of both arginine and ornithine it is capable of influencing the production of Growth Hormone.

References

1. *A Symposium on Amino Acid Metabolism*, Johns Hopkins Press, 1955.

Ornithine

Ornithine is not an EAA. It is a most important constituent of the urea cycle and is the precursor of other amino acids such as citrulline and glutamic acid, as well as proline.[1] Ornithine's therapeutic value lies in its involvement in the urea cycle, and in its ability to enhance liver function. It is used in the treatment of hepatic coma states.[2]

Ornithine is formed when arginine is altered by the enzyme arginase. According to Pearson and Shaw in their controversial book, *Life Extension*[3] Growth Hormone is released in response to supplementation of 1 g to 2 g of

ornithine taken on an empty stomach at bedtime. It is also
claimed that the immune system is thus stimulated[4]
improving the immune response to bacteria, viral agents
and tumour activity. (See also Arginine, page 37).

Dr Jeffrey Bland comments on this type of approach
which he calls 'experimental pharmacology using nutritional factors', saying that, in effect, there is no way of
knowing what the long-term impact of such an approach
will be. There are no controls, and no follow-ups, and in
short the use of such methods, for more than the short term
(months only), is to be questioned. Caution should be
employed in the use of ornithine by anyone with a history of
schizophrenia, who may find a worsening of associated
symptoms if this or arginine is utilized excessively.

References
1. *Pharmazie*, Vol.15, pp618–622, 1960.
2. *Symposium on Amino Acids*, Johns Hopkins Press, 1955.
3. Pearson, D. and Shaw, S., *Life Extension*, Warner Books, 1982.
4. Mindell, Earl, *Ornithine*, pamphlet, 1982.
5. Personal Communication, 1984.

Serine

This is a hydroxy-amino acid. It has sugar producing
qualities and is very reactive in the body, taking part in
pyrimadine, purine, creatine and porphyrin synthesis. It
takes part in a reaction with homocysteine (which is derived
from methionine) to form cystine.[1]

Its main use is in cosmetics where it is added as a natural
moistening agent, involved in skin metabolism.[2]

References
1. Greenstein and Winitz, *Chemistry of the Amino Acids*, Wiley, New
   York, 1961.
2. Meister, A., *Biochemistry of Amino Acids*, Academic Press, New York,
   1965.

Glutathione *treatment of for wide range of degenerative diseases.*

Glutathione is a tripeptide (made of three amino acids) comprising the three amino acids cysteine, glutamic acid and glycine. The value of this biologically active compound is in the prevention and treatment of a wide range of degenerative diseases.

Its role as a deactivator of free radicals is well established.[1] Free radicals, often the result of peroxidised fats, are immune system suppressors, mutagens, carcinogens and encouragers of cross-linkage and thus the ageing process. Prevention and slowing of free radical activity is one of the major contributions of that class of substances which act as antioxidants such as vitamins A, C, E, the mineral selenium and amino acids such as methionine, cysteine and the compound amino acid, glutathione. Since free radicals comprise a separated part of a molecule, with one or more unpaired electrons, they are extremely reactive and can result in cellular damage when they unite with other molecules. Lipid peroxidation occurs when saturated or unsaturated fats are exposed to oxygen (rancidity). Peroxides result, and one such is hydrogen peroxide (bleach). Free radicals are part of the end result of this process. Interaction between free radicals and DNA and RNA can result in genetic alterations within the cell, resulting in biochemical anarchy. The interaction of free radicals with protein structures results in, among other things, the gradual development of cross links in collagen fibre, which is the characteristic sign of ageing. The tissues literally become constricted and tight, interfering with cellular circulation and drainage, and in texture become leathery, contracted and stiff. Glutathione is uniquely qualified to act against free radicals which produce this intensification of the ageing process.[2] This activity is conducted extracellularly by glutathione against free radical activity as well as lipid peroxides, which are deactivated. Intracellularly the activity of a glutathione-related enzyme, glutathione peroxidase, accomplishes the same task. In this enzyme glutathione is combined with selenium.

Trials at the Louisville School of Medicine have clearly

demonstrated the connection between ageing and the reduction in glutathione's presence. Comparing young and old animals it showed that glutathione was reduced in all tissues by as much as 34 per cent. Thus the ability to detoxify, as well as ageing through cross linkage of proteins, was markedly different in the older animals.

Glutathione has been shown in trials at Harvard Medical School to have the ability to enhance the immune protective status of certain cells. In trials in which cigarette smoke was introduced into a tissue culture, the usual result of impairment of phagocytic function was inhibited by glutathione.[3]

The possibility that there is a role for glutathione in cancer prevention comes from trials in which glutathione produced regression of induced liver tumours, when administered in late stages of tumour development.[4] In rats in which the cancer causing chemical would normally produce 100 per cent liver cancer development, there was a total of over 80 per cent alive and well after two years when glutathione was also administered.

Heavy metal detoxification is a further area in which glutathione has been useful.[5] It is effective in removing harmlessly from the body, lead, cadmium, mercury and aluminium.

Glutathione is found to be helpful in assisting the liver in its detoxification of liver peroxidation. Thus alcohol-produced damage of the liver is thought to be prevented in several ways by glutathione. In the first place there is actual reduction of hydroperoxides, prior to their attacking saturated fats, as well as the conversion of lipid hydro-peroxides into harmless hydroxy compounds. Glutathione also enables the liver to detoxify undesirable compounds for excretion, via the bile, through the action of glutathione-S transferases.[6]

Blechman and Kalita point out that, as with any substance within the body, whether this is a vitamin, a mineral, or anything else, it is necessary to determine specific bio-chemical needs, on an individual basis, and that what assists one person in the progression from ill health towards optimum health may not do so for another.[7] Glutathione appears to be a most promising, naturally occurring, compound with ramifications spreading throughout the processes of detoxification and ageing.

References

1. *Science*, 179;588–91, 1973.
2. *Physiology Review*, 48;311–73, 1968.
3. *Science*, 162, 810, 1968.
4. *Science*, 212, 541–2, 1980.
5. *J. Am. Med. Assoc.*, 187;358, 1964.
6. *Functions of Glutathione*, New York: Springer Verlag, 1978.
7. Kalita, D. and Blechman, S., *The Biochemical Powers of Glutathione*, pamphlet.

CHAPTER 7

Amino Acids in Action: the Power and the Potential

In this section, we shall look at those conditions where amino acids can normalize biochemical imbalances and thus facilitate healing. The replenishment of a deficiency, or rebalancing of an imbalanced ratio between nutrients, simply allows the normal functions of the body, which include its capacity for repair and healing, to operate more efficiently. In certain instances, though, specific pharmacological effects are achieved over and above the replacing of deficient nutrients. DLPA (D and L phenylalanine) or tryptophan is used this way, in pain control treatment. Where possible the text will note this difference in therapeutic use.

While the emphasis of this book is on the role of amino acids in healing, it should go without saying that many other factors are involved in ill health apart from amino acids. Amino acids should never be seen as the only way in which the body should receive assistance. Underlying causes should always be sought and removed or dealt with as appropriate. Nutrition is not the only factor causing disease and dysfunction, although it is one of the most potent influences, and has a part to play in almost every condition, since it can improve the underlying health of the individual and hence help speed recovery.

We need to pay attention to the individuals with the complaint, and all the variables which make them unique, including their nutritional needs, rather than just to the condition. Anyone who treats all headaches, or all stomach aches, or all asthma attacks, for example, with a standard approach is practising bad medicine, as these can all have different causes.

Why then have we, in this section, chosen to present the evidence for the power and potential of amino acids in a format which emphasizes the usefulness related to specific conditions; which does just what we suggest should not be done, and looks at particular health complaints, and what individual amino acids (and other nutrients) can do for them? The reason is simple.

The only other way of presenting evidence would be by describing a long series of individual case histories, showing how amino acids and other nutrients were selected in particular cases, to meet specific and unique needs, rather than dealing just with the symptoms (insomnia, hypertension etc). This would be a preferable way of assessing amino acids. However, were case histories of that sort to be used to illustrate the usefulness of amino acids in therapy, they would be labelled by detractors as being merely anecdotal evidence which 'proves nothing'. This is because of the current medical vogue which pays attention only to trials dealing with diseases or conditions rather than the requirements of individual patients. The various types of 'acceptable' study are explained below.

A group of people with a particular condition are all treated in the same way, and are then compared with another group of people with the same condition, who received dummy (placebo) medication. This is called a **placebo controlled trial**. The people receiving the dummy treatment (or no treatment at all in some instances) are called the **control** group. If the medical staff administering such a trial are unaware of which patients are receiving the real medicine, and which the dummy, the trial would be called a **double blind, placebo controlled trial**; double blind as neither the patient nor the doctor knows whether or not the patient is taking a medicine, nutrient or a dummy, until after the study is completed and benefits (if any) are analysed.

In some studies, the people taking the medicine (or nutrient) are switched half way through the trial, so that they then receive the placebo (and those on the placebo start to receive the 'real' medicine). This is then called a **double blind, placebo controlled, cross-over trial**. In yet another variation the patient is assigned to the placebo or the 'real medicine' group haphazardly, in a random manner, and this is then called a **'randomized' trial**. We might then end up with a **double blind**, **randomized**, **cross-over**, **placebo controlled trial**. At the end of such studies the results are analysed and the researchers decide whether or not they are statisically 'significant'.

These terms will be used in the text and should be understood to indicate the type of trial performed and whether or not the benefits were statistically better than those which might result from mere change. It should be noted that in some instances, placebos do better, statistically, than the drug being tested, even in some very serious conditions. The research studies discussed in the text will give the reader an understanding of the ways in which the evidence was gathered. Anyone who wishes to have a deeper understanding of the biochemistry of amino acids and the amazing complexity of their interactions with each other and with the other nutrients and enzymes which allow the body to function, should refer to the books mentioned in the text. The main objective in writing this book is to promote an understanding of the potential of amino acids as powerful aids to healing.

Alcohol damage and craving

One of the most serious side effects of excessive alcohol consumption is the damage to the liver. Recent studies have indicated that as little as a glass and a half of wine, or a pint of beer, is regarded as the maximum safe daily intake beyond which damage begins to occur.

Carnitine was studied in an animal trial in which rats were fed an enormous 36 per cent of their total calory intake as ethanol. They developed, not surprisingly, massive liver damage (hepatic steatosis) involving accumulation of fatty acids, cholesterol, phospholipids, triglycerides etc.

When they were given supplements of carnitine, as well as its precursors lysine and methionine, it was found that the carnitine supplemented rats developed significantly less fatty degeneration of the liver than those not supplemented. No advantage was noted in adding the lysine and methionine. The findings suggest that chronic alcoholics have a functional deficiency of carnitine.[1]

NOTE: Animal studies are considered by the author to be undesirable for many reasons, not least of which is the suffering inflicted on the unwilling participants. There is also the very real fact that species are different in the ways in which they react to substances and foods. The animal studies presented in this book should be viewed as merely supporting evidence and not as conclusive findings, which human studies might represent.

In another trial, the use of *mixed free form amino acids* was studied in the treatment of 35 patients with alcoholic hepatitis. Half (17) received a high protein (100 g daily), high calory (3000 kcal) diet and between 70 and 85 g of free form amino acids in a balanced mixture. The symptoms of those treated in this way were much improved after a month, with none of these seriously ill individuals dying. The other 18 patients with similar conditions received the same diet but no amino acids, and whilst there was some improvement in this group due to better dietary patterns than previously, it was not as great as in the group treated with amino acids, and four of these 'control' patients died in the month of the study.[2]

A study at Johns Hopkins University School of Medicine showed that patients with acute alcoholic hepatitis (liver disease) had depressed levels of most of the essential and non-essential amino acids at the time of hospitalization. A similar, but less pronounced, pattern of amino acid deficiency was noted in alcoholics without liver disease, very similar to those found in African or Asian famine victims, due to dietary protein deficiency.

When alcohol was stopped, a diet which included protein at a level of between 1 and 2 g of protein per kilogram of body weight daily was introduced. This pattern improved the amino acid status of the alcoholics with liver disease but failed to normalize such levels as compared with non-alcoholics, despite cessation of alcohol consumption and an adequate diet. It was found that despite simultaneous supplementation of vitamin B_6 (pyridoxine) the levels of its derivative known as PLP (pyridoxal-5'-phosphate) failed to normalize in these individuals. It is known that unless PLP levels approach normal, amino acid metabolism is altered and absorption and transportation of amino acids from the bowel are reduced.

This makes a strong case for supplementation with free form amino acids as opposed to a simple increase in protein intake, in individuals with severely compromised biochemistry due to alcohol abuse.

Supplementation of PLP intravenously until normal function is achieved may also be called for in such cases.[3]

Glutamine was found to *dramatically diminish the craving for alcohol* in nine out of 10 patients supplemented with 2 g daily of glutamine in divided doses.

The patients, their friends and families all stated that there was a reduction in the craving for alcohol as well as less anxiety and better sleep.

This was a cross-over study, and it was found that three of the patients continued to respond to the placebo after glutamine was stopped, but no patient who started on the placebo produced positive results.[4]

The distinguished researcher Professor Roger Williams states that this ability of glutamine to check a craving (e.g. for sugar or alcohol) probably relates to its effect on the appetite centre in the brain. He suggests an intake of 2 to 4 g daily.[5]

Drs Janice Phelps and Alan Nourse in their book *The Hidden Addiction*[6] present a programme for treatment of *alcohol addiction* and note that as part of the programme *free form amino acid complex (all the amino acids) should be given in doses of 500 to 1000 mg three times daily with vitamin B_6 (100 mg) to be taken an hour before meals on an empty stomach. They too recommend glutamine in a dose of 500 to 1000 mg four times daily between meals.*

Additional nutrients in their programme include vitamin C, 8 g daily; vitamin B_3 (niacinamide) 3 g; pantothenic acid (vitamin B_5), 1500 mg daily; vitamin B_6 with the amino acids and additionally as a diuretic (caution: doses of B_6 in the quantities which they recommend, in excess of a gram a day, *could produce neurological symptoms of peripheral neuritis).* They also prescribe adrenal cortex extract. In addition they suggest phenylalanine, tyrosine and tryptophan for symptoms of depression and insomnia which might accompany withdrawal from alcohol. These will be discussed when depression and insomnia are considered later in this section.

Dr Robert Erdmann and Meirion Jones in their book *The Amino Revolution*[7] outline a programme for *alcohol addiction* involving *complete free form amino acid, glutamine, glycine, tryptophan and phenylalanine. In addition co-factors such as vitamins B_3, B_{12}, C, zinc, selenium and what are termed catabolic amino acids, methionine, taurine and aspartic acid are recommended.*

This latter suggestion (catabolic amino acids) relates to different phases of the metabolic process in which energy release occurs in response to the breakdown of protein structures. Thus the time of day at which nutrients are taken, which affect these and other processes (which are tied to our 'body clocks'), is important. Catabolic nutrients should be taken between 4 and 10pm. For dosage recommendations and discussion of this concept (based on the work of a noted New York physician, Emanuel Revici MD) refer to the Erdmann/Jones book.

References

1. Sachan, D. et al. *Carnitine and alcohol induced fatty degeneration of the liver*, American Journal of Clinical Nutrition, 39:738–44, 1984.
2. Nasrallah, S. *Amino acid therapy in alcoholic hepatitis*, The Lancet, 2:1276–7, 1980.
3. Diehl, A. et al. *Plasma amino acids in alcoholics*, American Journal of Clinical Nutrition, 44:453–60, 1986.
4. Rogers, L. et al. Quarterly Journal of Studies on Alcohol, 18(4):581–7, 1957.
5. Williams, R. *Nutrition against Disease*, Bantam Books, 1981.
6. Phelps, J. and Nourse, A. *The Hidden Addiction*, Little Brown, 1986.
7. Erdmann, R. and Jones, M. *The Amino Revolution*, Century, 1987.

Amyotrophic Lateral Sclerosis (ALS)

ALS is a fatal wasting disease which has recently been shown to respond well to two separate amino acid strategies.

In the first at Mount Sinai Hospital, New York, in a pilot study, it was demonstrated to strongly improve extremity (arms and legs) muscle strength in people with ALS. Although, as yet, survival rates have not been shown to be increased, quality of life was certainly enhanced.[1] This was a placebo controlled study using the branched chain amino acids (BCAA) valine, leucine and isoleucine. One theory as to why BCAA helped involves their ability to activate glutamate dehydrogenase (GDH) which influences glutamate metabolism.

The researchers state:

> Our data showed that human brain GDH can be activated by physiological concentrations of L-leucine, suggesting that dietary supplementation with BCAA may provide a new therapeutic approach to human neuro-degenerative disorders in which modification of glutamate metabolism might prove beneficial.

Various other mechanisms are also possible for the noted improvement including the observation that because levels of both tryptophan and tyrosine in the brain are high in cases of ALS, and since BCAA lower such levels, this could influence neurological and brain function beneficially.[2] When there are high levels of tryptophan in the brain the neurotransmitter serotonin as well as tryptophan metabolites such as quinelonic acid are increased. This factor is known to be neurotoxic, negatively influencing glutanergic neurotransmission.

Clearly the precise mechanism for the usefulness of BCAA in ALS is not yet established. However it is safe and effective and well worth using. A second report involved the use of threonine in treating ALS.[3] 2 to 4 grams daily of L-threonine was given to 15 patients suffering from ALS.

Within 48 hours there were marked improvement in the levels of spasticity, lessened drooling and fasciculation as well as enhanced energy in seven patients, lesser but notable improvements of a similar nature were seen in three patients while five remained unchanged with this supplementation. The researchers consider that threonine increases levels of the neuro-inhibitory amino acid glycine in the cerebrospinal fluid, thus counteracting the known excess of excitatory amino acids in ALS patients. Threonine is low in grains and some vegetarians may be receiving inadequate levels of this amino acid.

References

1. Plaitakis, A. *et al.*, 'Branched chain amino acids in ALS' *The Lancet*, 7 May 1988, p.1015.
2. Correspondence, *The Lancet*, September 1988 pp.680–1.
3. Study of ALS Patients. *American Family Physician*, vol.37, No.6 p.312–1988.

About Aspartame (NutraSweet)

Amino acids are used in many commercial processes, a recent example being as artificial sweeteners for food.

Dr Michael Weiner, author of *Maximum Immunity*, discusses the harmful effects on the immune function of the amino acid combination marketed as the sweetener NutraSweet or aspartame.[1] Once ingested in a cold drink or artificially sweetened food, aspartame, Weiner informs us, breaks down into its constituents, the amino acids phenylalanine and aspartic acid, and the following sequence occurs. 'Methanol (wood alcohol) is formed. Many foods in nature contain methanol, including drinking alcohol, but most sources of methanol in nature are accompanied by ethanol, and it turns out that ethanol is a specific antidote for methanol.' Weiner points out that, when metabolized in the body, methanol (which remember is the end product of aspartame or NutraSweet) becomes the highly poisonous and immune suppressing substance formaldehyde (used for embalming bodies).

If methanol was accidentally consumed, the standard

procedure would be to pump the stomach and then to get the individual to consume a large amount of ethanol, making him or her more than slightly drunk. By saturating the system with ethanol in this way, methanol is degraded into acetaldehyde, resulting in drunkenness and a hangover – preferable to death which could result were the methanol allowed simply to degrade into formaldehyde. When NutraSweet is consumed in sweets, soft drinks etc there is unlikely to be any counterbalancing ethanol intake to allow the relatively safe degradation into acetaldehyde.

Thus, as Weiner points out, the only safe way to consume anything containing this undesirable sweetener would be to accompany it with an alcoholic beverage, not perhaps the best prospect for the health and immune system of a three-year-old who happened to be eating an ice-cream sweetened with aspartame!

Further dangers of the use of aspartame are highlighted by the internationally renowned researcher Professor Richard Wurtman of the Massachusetts Institute of Technology. In a report he shows that a number of neurochemical changes may result from its use, with serious potential consequences. In rats aspartame was shown to double the levels of phenylalanine in the brain, which effect was redoubled when carbohydrates (sugars) were consumed at the same time. This combination raised the levels of tyrosine (which derives from phenylalanine) in the brain by over 300 per cent! There was coincidental depression (by 50 per cent) of the normal increase in brain levels of tryptophan, which would usually follow ingestion of carbohydrates. The amount of sweetener used in this study was the equivalent to that consumed by a normal North American child on a hot afternoon (soft drinks, sweets, ices etc). The full implications of such effects on the brains of children were not clear at the time of the study, but subsequent correspondence from Professor Wurtman on the subject indicates that the anxiety felt by many was justified.[2]

Writing to *The Lancet*, Professor Wurtman describes the possibility of a link between seizures (fits) in healthy adults and the use of aspartame. He describes three cases in which the association is assumed. In one a 42-year-old woman drank 3¾ litres of diet soda daily. She experienced mood swings, depression and headaches, along with nausea. Ultimately she had seizures (epilepsy).

A second case involved a 27-year-old male who drank four or five glasses of diet (non-sugar and sweetened with aspartame) drink daily. He developed twitches at night along with abnormal breathing, a severe headache and eventually grand mal seizures (epileptic fits). The third case involved a 36-year-old man who drank nearly a litre daily of aspartame-sweetened tea. He too developed seizures. *In all cases headaches and other symptoms disappeared after aspartame sweetened drinks were stopped.*

In his letter, Wurtman described a sequence of events as involving increased levels of phenylalanine, leading to abnormal levels of catocholamine and serotonin production in the brain, due to imbalances caused by the absence of other neutral amino acids. This would set the scene for the sort of symptoms listed above.[3]

It is as well to consider that in certain instances the use of supplemental amino acids could cause imbalances similar to this, unless strict attention is paid to the guidelines given as to dosage etc. Because something is helpful it does not mean that more of the same will be better. This is especially true of some of the amino acids which sometimes work therapeutically in small amounts, and not in large ones. If advice given in this book is followed, then there will be no such dangers. Please follow the guidelines and doses recommended.

References

1. Weiner, M. *Maximum Immunity*, Gateway Books, Bath, 1987.
2. Wurtman, R. New England Journal of Medicine, 309:249, 1983.
3. Wurtman, R. Letter to *The Lancet*, 8463:1060, 1985.

Atherosclerosis and cardiovascular/circulatory problems

There is no suggestion intended that amino acids alone are the answer to these conditions. However, they form a major element in the pathology of damaged cardiovascular structures, and can be important in their healing. Cardio-vascular and circulatory problems kill more people than any other disease in industrialized society, and have many dietary causes including *excessive intake* of hydrogenated (saturated) fats, mainly of animal origin, excess intake of refined carbohydrate, excess cholesterol levels in the sys-tem, *low levels* of selenium, vitamin C, specific essential fatty acids (such as omega 3 and omega 6), potassium, chromium, vitamin E etc.

A diet low in animal fat, cholesterol, alcohol, caffeine and refined carbohydrate, high in complex carbohydrates (vege-tables, fruits, beans, wholegrains etc) and supplemented as indicated with nutrients, especially vitamins B, C and E, the essential fatty acids, and calcium, magnesium, chromium, selenium, zinc, coenzyme Q_{10} etc, will help to enhance cardiovascular function if used alongside the amino acids discussed below. Herbal products such as ginger and garlic are also noted as being useful in supplying essential nutrients and having beneficial actions on the circulatory function. Smoking should of course be avoided.[1,2,3]

A human study showed that individuals consuming a diet high in animal protein, whose blood levels of the amino acids *lysine* were in a ratio of 3.5 to 1 or higher in relation to *arginine* were at much *higher risk of arteriosclerosis* due to excessive levels of *lysine*. Vegetarians have been shown to have a much lower ratio of these two amino acids than meat eaters, and consequently a lower incidence of arterio-sclerosis. The ratio of lysine to arginine in meat is between 3 and 4 to 1, whereas in plant proteins it is between 1 and 1.25 to 1.[4]

In a double blind, cross-over study 44 men with *stable, chronic angina* which appeared after exercise were given either *L-carnitine* (1 g twice daily) or a placebo. After a month, it was found that the patients receiving carnitine could take more exercise without inducing pain, whereas the placebo group remained unaltered.[5]

Italian research has shown that during *acute or chronic cardiac ischemia* (lack of oxygen reaching the heart muscle, often due to arteriosclerosis) there was both a deficiency of *carnitine* in these tissues and an accumulation of various fatty substances which carnitine is able to deactivate. Patients who had died of *acute myocardial infarction* were shown to have gross deficiency of *carnitine* in the heart muscles. It was suggested by the researchers that were adequate carnitine available in the body, the tissues could have been restored to normal.[6]

One Japanese study involved 62 patients suffering from *congestive heart failure*. This was a double blind, randomized, cross-over, placebo controlled trial, in which the amino acid *taurine* was supplied in three daily doses of 2 g each for four weeks. After four weeks the patients receiving taurine showed many *significant improvements* including better breathing, fewer palpitations, less swelling (oedema), improved laboratory and X-ray evidence as to heart status, and an overall improvement in function (by the New York Heart Association functional classification) which simply meant that the patients receiving taurine could do more than previously than the placebo group.

It was also found that patients receiving taurine could reduce their medication intake (digitalis and/or diuretics).

Neither heart rate nor blood pressure was affected by taurine intake. Taurine is known to be the most abundant free amino acid found in healthy cardiac tissues. The researchers concluded that taurine may be a useful agent that could be given by itself, or along with more conventional forms of therapy, for the treatment of congestive heart failure.[7]

In a study of 97 consecutive patients with *acute chest pains* it was found that those patients suffering *myocardial infarction (heart attacks)* had much higher levels of *taurine* being selectively 'leaked' into the bloodstream from the heart muscles. The requirement for taurine at such a time is very great and the normal supply became depleted, leading to its withdrawal.

The researchers noted that it acts in the same way as magnesium, having a *direct effect on the levels of potassium which pass into and out of the heart muscle cells*, a most important element in its health.[8,9]

The importance of *taurine* in *maintaining normal electrical and mechanical activity of the heart muscle* is shown in animal studies involving rats and guinea pigs. The electrolytic state of the heart was normalized by use of taurine when artificial damage was created in these animals by a variety of means.[10,11]

A study of the effectiveness of *tryptophan* in the treatment of *cardiac conditions* revealed that its ability to produce a relaxation of muscle tone *lowered the incidence of anticipated deaths from heart attacks by 15 per cent.* This effect was achieved by reducing the likelihood of heart spasm, racing heart rate and fibrillation (uncontrolled fluttering of the heart muscles).[12]

An animal study (using rabbits and rats) was conducted using *glycine, arginine and alanine* to see the effects on *reduction of increased levels of cholesterol.* These were artificially induced through dietary manipulation. It was found that arginine and alanine supplementation reduced serum cholesterol of rabbits by 20 per cent. When glycine was added as well, levels of cholesterol were halved.

The size and number of atherosclerotic lesions in the animals' aortas were reduced as the cholesterol levels came down with amino acid supplementation. The serum levels of cholesterol of rats were reduced by half when glycine was added. When casein (milk solids) was replaced with soy protein, cholesterol levels were reduced by two thirds. Excessive intake of casein also results in an imbalance in amino acid levels, which is beneficially affected by glycine and possibly arginine and alanine. The human implications of this study are not immediately clear except that a great many people ingest heroic amounts of dairy produce, which is known to have negative effects on atherosclerosis.[13]

In addition to the specific effects of amino acids reported above, methionine may be useful (in combination with an appropriate cholesterol reducing diet) in preventing arteriosclerosis (hardening of the arteries). This sulphur rich amino acid, vital for the production of certain enzymes which *protect the arterial wall*, also acts as a *detoxifier of heavy metals* (via a physiological method called chelation in which it literally grabs the metal atoms and removes them from the scene). It is also a *powerful antioxidant*.

A number of other nutrient co-factors are required to act with methionine for it to work efficiently, including vitamin B_6 (pyridoxine). We should also recall (see Chapter 8) that methionine (or lysine) becomes carnitine in the presence of adequate vitamin C.

Colin Goodliffe recommends a dosage of 600 to 2000 mg daily of carnitine to raise levels of beneficial lipoproteins and to lower cholesterol levels. [14] *Robert Erdmann and Meirion Jones maintain that a cocktail of nutrients should be employed to achieve maximum heart health, both in a protective and therapeutic setting. These include individual amounts of methionine, serine, tryptophan and histidine as well as a complete blend of free form amino acids. Co-factors such as vitamins B_3 (niacin), B_5 (pantothenic acid), B_6 (pyridoxine), B_{12} (cynaocobalamine) and folic acid, vitamins C and E, magnesium and zinc are also recommended. For the rationale of this approach read* The Amino Revolution *by Robert Erdmann PhD and Meirion Jones.*

References

1. Davies, S. and Stewart, A. *Nutritional Medicine*, Pan Books, 1987.
2. Werbach, M. *Nutritional Influences on Illness*, Thorsons, 1989; Third Line Press, California, 1987.
3. Goodliffe, C. *How to Avoid Heart Disease*, Blandford Press, Poole, 1987.
4. Sanchez, A. *Nutrition Reports International*, 28:497, 1983.
5. Cherchi, A. et al. *Carnitine effect on exercise tolerance in chronic stable angina*, International Journal of Clinical Pharmacology and Therapeutic Toxicology, 23(10):569-72, 1985.
6. *The Lancet*, Vol.1:1419-20, 1982.
7. Azuma, J. et al. *Taurine and congestive heart failure*, Circulation Research, 34(4):543-57, 1983.
8. Lombardini, J. et al. *Elevated blood taurine levels in acute myocardial infarction*, Journal of Laboratory and Clinical Medicine, 98(6):849-59, 1981.
9. Chazov, E. et al. *Taurine and electrical activity of heart*, Circulation Research, 35 (suppl) 3:11-21, 1974.
10. Sustova, T. et al. *Effects of Taurine on potassium, calcium, and sodium levels in the blood and tissues of rats*, Jn Vopr. Med. Khim., 32(4):113-16, 1986.
11. Franconi, F. et al. *Protective effect of taurine on hypoxia*, Biochemical Pharmacology, 34(15):2611-15, 1985.
12. Sved, A. et al. *Studies of anti-hypertensive action of L tryptophan*, Journal of Pharmacological and Experimental Therapeutics, 221:329-33, 1982.

13. Katan, M. et al. *Reduction in casein-induced hypercholesterolaemia and atherosclerosis in rabbits and rats by dietary glycine, arginine and alanine,* Atherosclerosist, 43:381, 1982.
14. Goodliffe, C. *How to Avoid Heart Disease,* Blandford Press, Poole, 1987.

Behaviour modification for aggression

As neurotransmitters are heavily involved in brain and nervous system function, and as amino acids are essential to neurotransmitters (see page 27), then it follows that behaviour might be modified when these are imbalanced, either in excess or deficiency. There is abundant evidence linking aggressive, anti-social behaviour with nutrient imbalances and with heavy metal toxicity (lead, mercury, cadmium, aluminium etc).

Alexander Schauss has shown that nutritional modification can alter aggressive juvenile delinquent behaviour dramatically. Among his findings were that many aggressive individuals were sensitive to specific foods, most notably dairy produce, and that removal of the offending food improved behaviour. In terms of toxicity he notes that the amino acids methionine, cysteine and cystine all contain sulphur compounds which naturally latch onto (chelate) toxic metals in the body and help in their removal (vitamin C, calcium, zinc and pectin also help in this process).[1]

Dr Stephen Levine, director of research for the Allergy Research Group, states that there is now considerable interest in how imbalances between two or more neuroregulators may cause behavioural disorders. There has been some research into the relationship between inadequate serotonin (derived from tryptophan in the body) and aggressive behaviour. Studies have reported that reduced levels of serotonin in the brain result in aggressive behaviour, and that low levels of norepinephrine and high levels of dopamine (both of which derive from phenylalanine and tyrosine) increase aggressive behaviour.[2]

It has also been shown that when there is a lot of

tryptophan in the diet, and the balance is high, compared with other amino acids, feelings of fatigue and inertia are reported.[3]

Recent experiments on rats have shown the link between aggression and tryptophan. When tryptophan was deliberately eliminated from the diet of these animals they became extremely aggressive and murdered each other (this is known as muricide in scientific jargon). This behaviour was stopped when tryptophan was added to the diet.[4]

As a result of this type of research tryptophan has been used on aggressive schizophrenic patients. Female schizophrenics have been shown to be low in tryptophan and their levels have increased with recovery.[5]

In an experimental double blind cross-over study, 12 male patients alternately received either placebo or a dosage of tryptophan. The patients with a large number of aggressive incidents improved on tryptophan. Their hostility also lessened and their mood improved. Those patients who had no such aggressive history worsened on tryptophan supplementation.[6]

References

1. Schauss, A. *Diet, Crime and Delinquency*, Parker House, California, 1980.
2. Levine, S. *Behavioural Ecology*, Bioscience, Vol. 2 Nos. 5 & 6.
3. Wurtman, R. *The Lancet*, 21 May 1983, p.1145.
4. Broderick, P., Lynch, V. *Behavioural and biochemical changes induced by lithium and tryptophan*, Neuropharmacology, 21:6671, 1982.
5. Gilmour, D. et al. Biological Psychiatry, 6:119, 1973.
6. Morand, C. et al. Biological Psychiatry, 18:575–8, 1983.

Candida Albicans (yeast overgrowth)

The health problems associated with yeast overgrowth are many and varied and range from genito-urinary disorders to digestive bloating and emotional instability, including depression, anxiety and phobic behaviour (see my book, *Candida Albicans, Could Yeast be Your Problem?* Thorsons, 1991).

The increase in Candida activity in recent years has been ascribed to a number of factors: increased use of antibiotics (which kill friendly bacteria in the gut, a normal part of the controlling mechanism against yeast spread); use of hormone drugs such as the contraceptive pill, which enhances yeast activity; and a dietary pattern rich in sugars, on which yeast thrives.

A variety of nutritional controlling strategies have been devised, including a low sugar diet coupled with the use of high-potency acidophilus cultures (superdophilus etc) to repopulate the bowel with friendly bacteria, the taking of garlic, olive oil (for oleic acid), the B vitamin biotin as well as aloe vera juice, and the trace element germanium, all of which have anti-candida potential. Specific anti-candida drugs, such as nystatin and caprystatin (a derivative of coconut), are often used.

Because of malabsorption and maldigestion problems, which are often associated with Candida activity, the use of a full range of free form amino acids is often suggested to enhance overall nutritional balance.

One of the problems in diagnosing Candida activity has been the fact that every man, woman and child in the world is host to this yeast, and it is therefore sometimes hard to know whether or not it is out of control and causing the types of symptoms described above. When Candida is out of control and rampant, amino acid profiles have, however, been shown to have distinctive patterns. When Candida is active, the following substances are usually found to be present in relatively large amounts in the urine: *phosphoserine, beta-alanine, gamma-aminobutyric acid, hydroxylysine, ornithine, anserine, carnosine,* phosphoethanolamine, ethanolamine, 1-methylhistidine, 3-methylhistidine, ammonia. The first seven of these (in italics) are not usually even detectable in urine (anserine is not even found in human tissues) and their presence indicates that they are the result of unusual activity in the body, or that they are by-products of yeast activity.

These findings were compared with profiles of other patients, not affected by Candida, and were found to be distinctive. Very low levels of arginine were measured in 96 per cent of Candida patients. The researcher in this study states that, 'These high levels of amino acids occur simultaneously and suggest metabolic disorders involving phospholipid metabolism, liver, kidney, maldigestion, malabsorption, connective tissue, and pyridoxal-5'-phosphate (vitamin B_6) metabolism.'[1]

What is fascinating is that when Candida is treated successfully by the methods outlined above, the changing pattern can be read on subsequent amino acid profiles, giving the doctor an objective method of assessing progress. Amino acid profiles are available in the UK and USA.

References

1. Traister, J. *Abnormal aminoaciduria pattern diagnosed with candidiasis*, Paper presented at meeting of American Society of Clinical Pathologists, Orlando, Florida 27 Sept 1986.

Cataracts

Over half the population over the age of 65 in the USA are affected by cataracts, and these are a major cause of blindness in Africa. They occur when ultra-violet light produces the development of free radicals which damage the lens.

A number of connections have been made between the development of cataracts and diet. Among the main causes are excessive sugar intake and high levels of dairy produce (mainly milk), as well as deficiencies (and excess) of the B vitamin riboflavin (B_2). It has been suggested that people with cataracts should have no more than 10 mg daily of riboflavin in supplement form. Zinc may also be deficient.

The antioxidant nutrients vitamin C (which should be present at 30 to 50 times its normal level in the bloodstream), vitamin E, and the mineral selenium are found to be helpful in retarding the damage to the eye caused by what are called 'free radicals'. These are minute destructive biochemical entities which occur when molecules with unpaired electrons cause a chain reaction of cell damage, leading, depending upon location, to changes such as the onset of arterial disease or even cancer, and in this case to cataracts. Antioxidants prevent this damage.[1]

Amino acids contain a number of important antioxidants, and those which are particularly effective in cataract treatment are cysteine and methionine levels of which decrease with age. Plentiful supply of both of these are essential for the synthesis of glutathione peroxidase, the main defender of the eye against the sort of damage induced by free radicals. Selenium is also important in this process.

An article in the *Journal of the American Medical Association* discusses the relationship between cataracts and a lack of glutathione peroxidase, and suggests that decline in glutathione peroxidase can be halted by diets rich in cysteine and methionine.[2]

Dr Alex Duarte has traced the mechanism of lens physiology, and has researched into the nutrients involved in cataract development. He points out that the enzymes hexokinase and phosphofructokinase are responsible for the energy production which keeps the so-called 'lens cation pump' mechanism active. This pumps sodium out and potassium in to the lens, thus maintaining high levels of amino acids for the synthesis of protein. If the energy supply becomes diminished, this pump mechanism slowly fails, resulting in loss of protein synthesis, loss of transparency, and hardening of the lens proteins, which change from being 80 per cent soluble to being only 35 per cent soluble in a lens with a cataract. Duarte also points to glutathione peroxidase as a major form of protection against this process, and suggests that treatment consisting of a combination of amino acids, enzymes and vitamins could correct the various imbalances.

This approach is claimed to be 80 per cent effective in slowing or stopping progression of senile cataracts. Trials in Bonn have demonstrated the safety and efficacy of this method through both double blind trials and animal studies. Of interest is the fact that it is normal for the concentration of vitamin C in the lens to be 30 to 50 times the level found in the bloodstream. This is vital for protecting the lens.

The aim in treatment is not to restore damaged tissues but

to normalize energy production in the eye and thus stabilize the operating mechanisms, improve protein synthesis and eventually retard the degenerative process.[3]

Cysteine and methionine are sulphur rich amino acids and are found in eggs, beans, garlic and onions. However, since it takes a great deal of any of these foods to make a significant contribution to the body's levels of amino acids it is suggested that supplements should also be taken.

Dosage for methionine is between 200 and 1000 mg daily, with water, away from mealtimes (no closer than 90 minutes to any protein intake or absorption of the amino acid will be impared).

Dosage for cysteine is between 1 and 2 g daily with vitamin C at the same time at a level of three times the amount of C to cysteine. Cysteine should also be taken away from mealtimes with water.

NOTE: Cysteine should not be given to diabetics without professional guidance.

References

1. Werbach, M. *Nutritional Influences On Illness*, Thorsons, 1989; Third Line Press, California, 1987.
2. Cole, H. *Enzyme activity may hold key to cataract activity*, Journal of the American Medical Association 254(8):1008, 1985.
3. Duarte, A. *Cataracts can be stopped without surgery*, Health Consciousness 5(4), 1984.

Chronic disease (mental or physical)

William Philpott MD has conducted extensive research into the deficiencies that cause chronic ill health. He makes the following claim: 'Our statistical study reveals that the majority of degenerative illnesses, whether physical or mental, have a vitamin B_6 utilization disorder ... it seems evident that one of the problems producing B_6 utilization disorder is low cystine (cysteine) ... Cystine is a necessary amino acid which the body makes from the essential amino acid methionine ... the universality of cystine not being formed from methionine in chronic degenerative diseases suggests that a state of addiction or toxins associated with addiction, interferes with methionine metabolism.'

Philpott's answer to this is to supplement all cases of chronic degenerative disease with cystine or cysteine at doses of 1½ g, three times daily for a month, with dosage then reduced to twice daily. In addition he suggests 50 mg of pyridoxal-5'-phosphate (the active form of vitamin B_6) three times daily. Philpott further points out that the most commonly deficient substance noted in amino acid profiles of people with degenerative diseases is the precursor of glutamic acid, alpha-ketoglutaric acid. This is associated with vitamin B_6 in many enzyme activities. Glutamic acid is also shown to be invariably low in these very sick people.

Philpott notes that the major body cycles of detoxification and energy production are linked by the amino acid, aspartic acid. In a complex set of biochemical steps he finds that supplementation of aspartic acid and citric acid helps to normalize the cycles and restore balance to many of these imbalances.

Dosages of 500 mg aspartic acid and 750 mg citric acid are suggested five times daily (on rising, at each meal and on retiring). If resulting digestive discomfort is noted, he recommends a quarter teaspoon of bicarbonate of soda as well.[1]

References
1. Information sheets, Philpott Medical Center, Oklahoma City, 1983.

Depression

There are many causes of depression. Many cases have been shown to relate to alterations in the body's biochemistry, although not in any uniform pattern. The disorder appears either as depression alone (uni-polar) or (far more seriously) in the form of mood swings from manic behaviour to deep depression (bi-polar). Manic behaviour involves violent, often aggressive physical activity, restlessness, and the feeling of being mentally supercharged. Depression is characterized by exhaustion and a withdrawal from any form of activity and involvement.

These conditions are known as affective disorders, and the biochemical alterations which take place may relate to

specific deficiencies or to excess amounts of toxic substances including heavy metals (eg vanadium) and sometimes of normally safe nutrients. For example, cases are
reported of mania (insomnia, delusions and excessive
energy etc) developing when large quantities of glutamine
(2 to 4 g daily) were ingested.[1]

In one study of 40 patients with *major depression* of whom
eight were bi-polar depressives, *L-phenylalanine (LPA)* (the
precursor of tyrosine) was given. Dosage was *500 mg in the
morning and again at noon, plus vitamin B_6* (100 mg twice
daily), for a week at the outset. Phenylalanine was then
*increased by 500 mg daily until benefits (or side effects) were
noted*, with a maximum intake of 4 g daily.

Thirty-one of the patients *improved almost immediately*,
including seven of the eight bi-polar depressives. It was
suggested that most of the others who found benefit were
marginally tending towards bi-polar symptoms and that
phenylalanine was more beneficial in such cases. Ten of the
patients were completely free of depressive symptoms after
supplementation.

Side effects of slight headache of short duration, con-
stipation, slight nausea and insomnia were noted in some
patients.[2,3]

In treatment *D-phenylalanine* was also effective in
relieving depressive symptoms in cases of plain depression
without manic tendencies.[4]

In a double blind controlled study 14 *depressed* patients
were given *DL-phenylalanine (DLPA)* in doses of *150 mg to
200 mg daily* while another group of patients (controls)
received an anti-depressive drug (imipramine). At the end
of the month's study no difference could be noted between
the two groups, showing that *DLPA was at least as effective as
the drug*.[5]

Arnold Fox MD, in his book *DLPA: The Natural Pain Killer
and Anti-Depressant* takes a strongly nutritional approach
to the care of depression, including supplementing
DL-phenylalanine and tryptophan.

He sees the effects of DLPA on depression to be the result
of three factors:

1. Increased levels of norepinephrine which relieves depression (the 'D' form appears the most likely cause here as it is less likely to be harnessed by the body for other uses than is the 'L' form).
2. Increased levels of phenylethylamine (PEA) which is a neurotransmitter closely linked to norepinephrine. Depressed individuals have very low levels of PEA, which antidepressive drugs tend to raise.
3. Increased manufacture of endorphins (literally *endogenous* (self-produced) *morphines*) which act to produce feelings of euphoria.

Apart from a range of vitamins and minerals designed to enhance general function, Fox suggests *500 mg of tryptophan twice daily, at 8am and 8pm, as well as 375 mg of DLPA with breakfast and with lunch. This dosage should be increased to three times daily after a few days. It is to be assumed that these should be taken well before meals rather than with food, although this is not Fox's advice, which goes against the trend of most.*

A variety of studies have been conducted treating both bi-polar and uni-polar disorders with the amino acid tryptophan either alone or in combination with other nutrients.

In an experimental placebo controlled, double blind trial 24 patients with *manic disorders* were treated for seven days with *12 g of tryptophan* daily (this was given in divided doses, without any form of protein 90 minutes before or after it and with a little sugar to enhance its absorption). After a week, half the patients were randomly selected to continue this dosage while the other half were given a placebo. In the second week, *only the patients receiving placebo were seen to display manic symptoms.*[6]

Tryptophan is more effective if also taken with vitamin B_6 (pyridoxine).[7] In uni-polar depression *doses of 4 to 6 g of tryptophan daily* (divided doses away from protein and with sugar) are seen to be more helpful than higher or lower doses.[8]

Studies show that *depressive patients* have *low levels of tryptophan* in their bloodstreams compared with normal individuals. In one, 50 depressive women with uni-polar depression had lower than average levels of tryptophan.

Patients who had recovered from depression also had lower levels than control patients, and those who had recovered without the use of antidepressive drugs had higher, more normal, levels than those who had been treated with drugs. A second study showed that amongst women suffering from post-natal depression, those with the deepest depression were the ones with the lowest levels of tryptophan in their bloodstreams.[9,10]

NOTE: Tryptophan is dangerous if taken at the same time as monoamine oxidase inhibitor drugs (MAO inhibitors) or tricyclics, which are often prescribed to depressed patients because it enhances their side effects. If you are on these, do not take tryptophan without first seeking professional advice.

The amino acid *tyrosine* which derives from phenylalanine has been shown to be *effective in treatment of some cases of unipolar depression*, taken in doses of 2 g, three times daily. Three out of five such cases, none of whom was receiving other medication, showed a *50 per cent reduction in levels of depression* in a four-week double blind placebo-controlled study. Levels of tyrosine rose in those patients whose depression was observed to improve. The only side effects noted were mild gastric upsets when tyrosine was taken without food.[11]

The trials and studies reported above indicate that *some* patients respond very well to particular amino acid supplementation when depressed or in a state of manic depression. In another study of severely depressed people, 80 per cent experienced relief after taking 100 to 500 mg of L-phenylalanine (which is converted into tyrosine in the liver) daily for a fortnight.[12]

The questions should be addressed, if phenylalanine is as helpful as this, why did the other 20 per cent fail to respond? Is it possible to identify, in advance, which patients will respond beneficially? Much of the answer has been provided by an Australian researcher, Dr Robert Buist. He has observed that some depressed people have low levels of the hormone noradrenaline in the brain (identified by the low levels in their urine of certain biochemical waste products such as MHPG or 3-methoxy-4-hydroxyphenthylene glycol). These people will have responded well to treatment with tricyclic drugs such as imipramine and desipramine, but will have a poor response to amitriptyline. They will also usually respond well to tyrosine supplementation since it is a precursor of noradrenaline. A second group who are biochemically different show normal or high levels of MHPG in their urine indicating high levels of noradrenaline in the brain. They would respond poorly to imipramine (tricyclic drugs) and well to amitriptyline, which tends to raise serotonin levels which are low in such cases. It is therefore possible to know in advance which amino acid will help in particular cases of depression by analysing the urine for MHPG, or by studying previous response to antidepressive drugs. This does of course require expert advice and assessment, and it is unwise to try and diagnose yourself.[3]

It is not surprising that precursors of neurotransmitters, such as the amino acids described in the studies above, should be so helpful. What is amazing is that medical science has taken so long to realize the potential of these amazing nutrients.

A neurotransmitter is a chemical messenger which enhances or retards transmission of nerve impulses as necessary. There are many different neurotransmitters, one of which is serotonin which derives from tryptophan. Others include dopamine and adrenaline which derive from tyrosine and phenylalanine. They are also important in weight control, as will be discussed later in this section.

The levels of serotonin will alter according to the supply of tryptophan or dopamine in the brain, with profound influence on mood and behaviour. High levels of ammonia in the brain affect mood and behaviour and can be removed by supplementation by glutamine.

Other nutrients which can be helpful in cases of depression include most of the B-complex vitamins, vitamin C, calcium, magnesium, zinc, iron, potassium and essential fatty acids (evening primrose oil or vitamin F). Food allergies or sensitivities should also be considered as causes and identified and eliminated.

This may call for periods on a severely restricted diet (the so-called Stone Age diet) and rotation diets which carefully programme consumption of specific food families until culprit foods are identified. Doctors who specialize in this form of treatment are called clinical ecologists.

References

1. Letter to the American Journal of Psychiatry 141, 10 October 1984.
2. Sabelli, H. et al. *Clinical studies on the phenylalanine hypothesis of affective disorder*, Journal of Clinical Psychiatry, 47(20):66–70, 1986.
3. *Phenylalanine: A psychoactive nutrient for some depressives?* Medical World News, 27 October 1983.
4. Beckmann, H. *Phenylalanine in affective disorders*, Advanced Biological Psychiatry, 10:137–47, 1983.
5. Beckmann, H. *DLPA versus imipramine*, Arch Psychiatr, Nervenkr, 27:58, 1979.
6. Chouinard, G. et al. *A controlled clinical trial of L-tryptophan in acute mania*, Biological Psychiatry, 20:546–57, 1985.
7. Green, A. et al. *Pharmokinetics of tryptophan*, Advanced Biological Psychiatry, 10:67–81, 1983.
8. Wallinder, J. Advanced Biological Psychiatry, 10:82–93, 1983.
9. Coppen, A. *Tryptophan and depressive illness*, Psychological Medicine, 8:49–57, 1978.
10. Stein, G. et al. *Relationship between mood disturbances and tryptophan levels in post-partum women*, British Medical Journal, 2:457, 1976.
11. Gibson, G. *Tyrosine for treatment of depression*, American Journal of Psychiatry, 147:662, 1980.
12. Gelenberg, A. *Tyrosine for the treatment of depression*, American Journal of Psychiatry, 147:622, 1980.
13. Buist, R. *Therapeutic predictability of tryptophan and tyrosine in treatment of depression*, International Clinical Nutrition Review, 3(2):1–3, 1983.

Diabetes Mellitus

The amino acids *cysteine, glycine and glutamic acid* (as well as vitamin B_3 and the mineral chromium) make up what is known as the glucose tolerance factor (GTF) which is profoundly important in our handling of sugars because they can make the effects of insulin more powerful. This is of great importance to diabetics, and these amino acids as well as the chromium and vitamin B_3 may be supplemented, under supervision, thus meaning that intake of additional insulin can be reduced.

Diabetes is a serious condition and any attempt at self-treatment should involve great care. Carbohydrate and fat selection (basically low refined carbohydrate and fat intake and high complex (unrefined) carbohydrate intake) should be low, and *use should be made of a variety of nutrient aids including many of the B vitamins, vitamins C and E, calcium, zinc, magnesium, manganese, phosphorus, potassium, bioflavonoids, coenzyme Q_{10} and the amino acid complex glutathione.*[1]

Glutathione (which is made up of glutamic acid, cysteine and glycine) is suggested for *diabetics* due to its ability to reduce the levels of dehydroascorbic acid, the precursor of vitamin C in red blood cells. Dehydroascorbic acid is increased in diabetics, having, it is thought, harmful effects. Vitamin C is generally in short supply in the plasma of diabetics.[2]

References

1. Werbach, M. *Nutritional Influences on Illness*, Thorsons 1989; Third Line Press, California, 1987.
2. Banarjee, S. *Physiological role of dehydroascorbic acid*, Indian Journal of Physiology Pharmacology, 21(2):85–93, 1977.

Enlarged prostate gland (benign prostatic hyperplasia or BPH)

This condition affects roughly half the male population in industrialized societies, and appears to be largely preventable (and often successfully treatable) using a combination of nutritional approaches, including supplementation with zinc and a selection of amino acids.

The mixture of amino acids reported on in the trials discussed below (glutamic acid, alanine and glycine) were supplemented in two capsules of 1 g each (of the mixture) three times daily for a fortnight, followed by one capsule three times daily.

A study was carried out in which 17 men with *BPH* were supplemented with a mixture of *glutamic acid, alanine and glycine*. It was found that the *retention of urine* (a major problem in BPH) was reduced in eight cases, while the symptom of pain or difficulty in passing urine (*dysuria*) was reduced in 14 cases. All benefits were achieved with no side effects.[1]

In one double blind study, patients were either given the three amino acids as listed above, or glutamic acid and alanine, or glutamic acid alone. Those receiving *all three amino acids* reported *marked subjective improvements in symptoms* as compared with the two control groups. There was however little palpable change in the size of the enlarged prostate noted in any of the patients, with any of the three forms of supplementation.[2]

In a controlled study of 45 patients the *triple amino acid mixture* was supplemented and resulted in 95 per cent of the patients reporting that the symptom of having to pass water during the night (*nocturia*) *was relieved*; 81 per cent reported *reduced urgency*; 73 per cent reported *reduction in frequency* of urination, while the problem of a long *delay in commencing the flow* when trying to pass water *was relieved* in 70 per cent of cases.[3]

In most cases of BPH, zinc supplementation has been shown to reduce both symptoms and size of the enlarged gland. *Dosage recommended is between 50 and 150 mg daily of zinc gluconate, or if available zinc orotate or zinc picolinate, not taken at mealtimes. In order to avoid the problem of the supplemented zinc interfering with absorption of iron, iron rich foods should be eaten at mealtimes with ½ g to 1 g of vitamin C to enhance its absorption.* Essential fatty acids (also known as vitamin F) such as evening primrose oil or linseed oil have also been found helpful in treating BPH (dosage 1 to 1½ g daily).

WARNING: Other factors (including infection and sometimes malignancy) may enter into enlarged prostate conditions and expert advice should be sought so that an accurate diagnosis is made before self-treatment is attempted.

References

1. Aito, K. *Conservative treatment of prostatic hypertrophy*, Hinyokika Kiyo, 18(1):41–4, 1972.
2. Shimaya, M. *Double blind test of PPC for prostatic hyperplasia*, Hinyokika Kiyo, 16(5):231–6, 1970.
3. Dumrau, F. *BPH: Amino acid therapy for symptomatic relief*, American Journal of Geriatrics, 10:426–3, 1962.

Epilepsy

Epileptic seizures are essentially the result of aberrant neurological discharges in the brain, sometimes apparently due to, or associated with, particular nutrient deficiencies, including vitamin B_6 (pyridoxine), magnesium, manganese, zinc, etc.

It is not recommended that people suffering from epilepsy should treat themselves, but rather, that they should seek the advice of a physician who is aware of nutrient connections with the condition.

Several amino acids have been shown to be effective in the treatment of epilepsy, including taurine and dimethyl glycine.

Studies show low levels of *taurine* (which is a neuroinhibitory agent) *and glutamic acid*, as well as high levels of *glycine*, in those sites in the brain where maximum seizure activity is occurring.[1] *Taurine is therefore recommended in the treatment of epilepsy in doses of 500 mg three times daily, away from protein meals and with water.*[2] Animal studies on rats show that taurine controls experimentally induced seizures. This effect was powerful and of long duration.[3,4]

It appears that taurine works by normalizing the balance of other amino acids, which in epilepsy are thoroughly disordered. Epilepsy causes lowering of serum levels in over half the amino acids, whilst raising those of taurine, except in the cerebrospinal fluid which bathes the brain, where taurine levels are reduced during epileptic seizures. Zinc, which raises serum taurine levels, is low in epileptics.

Dimethyl glycine (DMG), which is formed in the body when homocysteine is converted into methionine before the formation of glycine, was observed to *reduce seizure frequency* in a male aged 22, who had a long history of generalized epileptic seizures. When supplementation of 90 mg twice daily of DMG was introduced, seizures were reduced from 16 to three a week. On each occasion that DMG was withdrawn the seizures recurred at their previously high level.[5]

NOTE: See also pages 95–97, for the link between epilepsy and the amino acid based artificial sweetener, aspartame.

References

1. Sherman, J. *Taurine in Nutrition*, Comprehensive Therapy, 35:672, 1979.
2. Werbach, M. *Nutritional Influences on Illness*, Thorsons, 1989; Third Line Press, California, 1987.
3. Mantovani, J. et al. *Effects of Taurine on Seizures*, Arch Neuro, 35:672, 1979.
4. Huxtable, R. et al. *Prolonged anticonvulsant action of taurine*, Canadian Journal of Neurological Science, 5:220, 1978.
5. Roach, E.N. *N-Dimethylglycine for epilepsy*, Letter to the New England Journal of Medicine, 307:1081-2, 1982.

Fatigue

There are many different causes of fatigue, which is after pain the commonest symptom reported to doctors. There is therefore no obvious simple prescription that can help all types of fatigue, which result from deficiency, toxicity, organic disease (diabetes, cardiac disease etc), chronic pain, obesity, psychosocial problems such as depression, chronic infections, inadequate exercise or sleep, environmental factors such as pollution, lifestyle factors such as smoking, alcohol, drug abuse etc, and many other causes including personality or genetic factors.

There are, however, certain nutritional supplements that can help in most cases of fatigue, including the B-complex vitamins, vitamin C, magnesium, potassium, zinc and a number of the amino acids as listed below. Many of these nutrients assist in the energy manufacturing and detoxification cycles of the body. Two recently researched nutrients, organic germanium and coenzyme Q_{10}, also have a specific relationship to energy. Books are available about fatigue which give information about the problem; we shall concentrate here on amino acids.

Aspartic acid is a major element in the energy cycle of the body and is therefore a useful supplement to relieve fatigue. Between 75 per cent and 91 per cent of some 3000 patients treated with *potassium/magnesium aspartate* (1 g of each twice daily) reported pronounced relief from fatigue, while fewer than 25 per cent of people in the control group receiving placebo reported increased energy. Benefits were normally reported within four days although in some cases it took as long as 10 days before benefits were noted. Treatment continued for four to six weeks, and in many instances fatigue did not recur when supplementation stopped. Dryness in the mouth and some gastrointestinal problems were sometimes noted with this treatment.[1]

In an experimental double blind study 87 out of 100 patients suffering from chronic fatigue reported marked improvement after five to six weeks of supplementation with aspartates.[2]

Dr Earl Mindell, author of *The Vitamin Bible*, maintains that aspartic acid increases resistance to fatigue, as well as stamina and endurance. It also detoxifies ammonia from the body, thus enhancing nerve function.[3]

Dr R Krakowitz recommends a full complement (balanced) of free form amino acids to counteract fatigue. Ten to 20 such capsules (containing between 500 and 800 mg of free form amino acids) are recommended two or three times daily for severe fatigue. This should be between meals with diluted fruit juice and never with a protein (milk for example) as this will delay or stop absorption. This approach should not be used as a long-term strategy but more as a short-term rebuilding approach for people recovering from severe fatigue of whatever cause.[4]

References

1. Gaby, A. *Aspartic acid and fatigue*, Curr. Nutr. Therapeut., November 1982.
2. Formica, P. *The Housewife Syndrome*, Current Therapeutic Research, 4:98, 1962.
3. Mindell, E. Information sheet on L-Aspartic acid, 1981.
4. Krakowitz, R. *High Energy*, Ballantine Books, New York, 1986.

Gallbladder disease

Research into the major nutritional causes of gallbladder disease has shown that a sugar rich diet is a major factor. Sufferers should reduce their intake of fats and of animal based foods, and eat foods rich in fibre (complex carbohydrates). Vegetarians are less likely to suffer from gallbladder disease than non-vegetarians. Women are more prone than men to problems in this area, especially if they are obese and have a high fat and sugar diet. A diet rich in polyunsaturated oils is also harmful: the safest oils are monounsaturated oils such as olive oil. Intolerance to certain foods as well as reduced levels of hydrochloric acid are not uncommonly associated with gallbladder problems. A variety of deficiencies have been noted accompanying gallbladder diseases, including vitamin C and vitamin E, which may beneficially be supplemented, as may essential fatty acids, lecithin and the amino acid taurine (see below).

Animal studies indicate that *taurine* has powerful preventive effects in *gallstone formation*. Three groups of mice were fed a cholesterol-free diet, a diet rich in the elements which help produce gallstones (cholesterol and sodium cholate) and no taurine, or a similar diet with taurine. *Only those mice receiving the gallstone-enhancing diet and no taurine actually developed stones.*[1]

Researchers into the use of taurine on guinea pigs have noted that 'taurine is rapidly emerging as one of the more interesting and ubiquitous amino acids'. Previous studies have shown that the majority of bile acids are joined by either glycine or taurine, making what are in fact natural detergents which increase the emulsification of fats, helping them to be more readily assimilated and easily metabolized. Those which are conjugated (joined) by taurine are the most useful to the bowel. This is influenced by many factors including age, diet, use of drugs, presence of hormones and any disease in the body. In the study in question bile acids were injected into guinea pigs. This would normally be expected to enhance development of gall-stones. When *taurine* was given in drinking water up to three days after or five days before the injections, *development of gallstones was prevented*. The conclusion of this study was that an increase in availability of taurine protects against the development of gallstones.[2]

References
1. Yamanaka, Y. et al. *Effects of dietary taurine on cholesterol gallstone formation*, Journal of Nutritional Science, Vitaminol, 31(2):226–32, 1985.

2. Dorvil, N. et al. *Taurine prevents cholestasis induced by lithocholic acid sulfate in guinea pigs*, American Journal of Clinical Nutrition, 37(23):221, 1983.

Herpes infection

Since the sexual revolution of the 1960s, herpes has been a major social disease. The discovery some years ago that amino acid manipulation can effectively reduce herpes virus activity and speed healing of active outbreaks has been one of the major phenomena of what Erdmann and Jones call the 'amino revolution'. Herpes attacks (involving HSV 11, or herpes simplex virus type 11) are characterized by the presence of clusters of clear fluid-filled vesicles on the genitalia or face, accompanied by severe pain and itching.

In some cases, there is no more than the occasional 'cold sore', whereas for others the disfigurement and discomfort can lead to major social and sexual problems. In the USA alone, some 300,000 new cases appear each year and roughly a third of the population (some 85 million people) have been exposed to what is known as venereal herpes at some time.

The amino acid strategy is discussed below, but other nutrients include zinc and vitamin C (with bioflavonoids) as well as topical application of vitamin E and zinc creams.[1]

The major finding relating to amino acids was that the *herpes virus thrived when there was a high level of arginine* (arginine added to a culture of herpes virus will speed reproduction of the virus) and was *inhibited when there was a high level of lysine.*

Once infected by the herpes virus an individual is unlikely ever to completely get rid of it. The virus remains dormant in the body after the initial infection until it is reactivated as the result of a stress factor such as sunburn, another infection (cold etc), menstruation, emotional upset etc. Stress changes the relative balance of the amino acids arginine and lysine in the circulation and this appears to be a critical factor in the re-emergence of the virus.

Diet also plays a major part so that it is important to

choose foods rich in lysine such as meat, potatoes, milk, brewers yeast, fish, chicken, beans and eggs, and to supplement with lysine as well as avoiding arginine rich foods (chocolate, peanuts, nuts, seeds and cereal grains).

It is thought that the reason why these two amino acids are important in herpes infections is because their chemical similarity means that the virus confuses them, incorporating lysine instead of arginine and thus depriving itself of its arginine supplies.[2]

One study showed that *lysine suppressed symptoms of herpes* in 96 per cent of 45 patients tested over a two-year period, with complete remission of herpes outbreaks and no side effects. The researchers reported that 'pain disappeared abruptly overnight in virtually every instance, new vesicles (blisters) failed to appear and resolution in the majority was considered to be more rapid than was their past experience. Patients were infection-free while on lysine, but found that within one to four weeks after stopping lysine, return could be predicted.' *The amount of lysine required to control herpes varied from case to case but a typical dose to maintain remission was 500 mg daily and active herpes required 1 to 6 g between meals to induce healing.*[3;4]

A study involving 41 patients indicated that with a daily dose of 1248 mg of lysine hydrochloride herpes outbreaks became less frequent and symptoms less severe. A lower dose of 624 mg failed to achieve this benefit, however.[5] Another study, this time a double blind cross-over trial involving 65 patients, showed that those taking lysine were far freer of recurrences than those taking a placebo.[6]

A combination of lysine and vitamin C together with bioflavonoids is used in the USA to combat the herpes virus, based on the known effects of lysine therapy and studies indicating the other two nutrients to be beneficial. Its brand name is Lysamin-C and each tablet contains 500 mg lysine, 100 mg vitamin C and 100 mg bioflavonoids.[7,8]

Robert Erdmann PhD has a complex approach to herpes, using lysine as well as a variety of other nutrients including other amino acids. He enhances physical resistance with a combination of tyrosine, DL-phenylalanine, glutamic acid and methionine, together with vitamins B_3 and B_6, C and magnesium, and mental tranquillity (a major element in recovery from, or control of, herpes infection) with tryptophan, histidine, taurine and glycine.

He adapts the work of the American Dr Emanuel Revici when treating the element of ongoing infection. This involves what are termed *the catabolic amino acids (in this instance) such as methionine, cysteine, taurine, aspartic acid and glutamic acid together with nutrients vitamin A, B₆, B₁₂, folic acid, and vitamin C (as calcium ascorbate) and magnesium aspartate, all taken in the evening to coincide with normal body cycles relating to stages of the metabolic process he wishes to influence.* See his book, *The Amino Revolution.*

CAUTION: An animal study on chicks indicated that prolonged use of lysine may stimulate the liver into producing excess levels of cholesterol, and this should be monitored.[9]

References

1. Werbach, M. *Nutritional Influences on Illness*, Thorsons, 1989; Third Line Press, California, 1987.
2. Passwater, R. *Nutrition and herpes*, Energy Medicine, 1(1):11, 1983.
3. Kagan, C. *Lysine therapy for herpes simplex*, The Lancet, 1:137 (26 Jan 1974).
4. Griffith, R. et al. *A multicentred study of lysine therapy in herpes simplex infection*, Dermatologica, 156:257–67, 1978.
5. McCune et al. Cutis, 34:366, 1984.
6. Milman, N. et al. *Lysine Prophylaxis in recurrent herpes simplex labialis*, Acta Dermatovener, 60:85–7, 1980.
7. Beladi, L. *Activity of some flavonoids against viruses*, Annals of the New York Academy of Science.
8. Leibovitz, B. et al. *Ascorbic acid Neutrophil Function and Immune Response*. International Journal of Nutrition Research, 48:159, 1978.
9. Schmeisser, D. et al. *Effect of excess dietary lysine may stimulate liver*, Journal of Nutrition, 113(9):1777–83, 1983.

High blood pressure (hypertension)

High blood pressure is a condition with varying causes, which means that there is no standard desirable treatment. It is the main cause of cardiac disease, which is the major killer in Western society. Anything which can naturally, and without side effects, reduce high blood pressure is of considerable potential value to mankind.

Bad nutrition is a common cause of hypertension as is low stress tolerance and lifestyle factors such as smoking, alcohol, and caffeine consumption (caffeine is found in tea, coffee, coke and chocolate). Nutritionally, a high fibre diet which is low in fat and sugar is the ideal pattern. Vegetarians tend to have lower blood pressure levels than meat eaters. To reduce hypertension, it is important to have a good supply of calcium, magnesium and potassium, and to keep sodium (salt) intake moderate to low.

Specific nutrients such as the essential fatty acids, known as omega-3, and coenzyme Q_{10}, are helpful in reducing hypertension, as is garlic. Stress reduction through relaxation methods or biofeedback techniques are helpful, as is regular exercise.

Amino acids can be useful in the treatment of hypertension in the following ways. For instance, tyrosine may be used, which is a derivative of phenylalanine and a precursor of substances known as catocholamines, a class of neurotransmitter, which includes dopamine and norepinephrine, the presence of both of which are known to decline with age. In studies at Massachusetts General Hospital, psychiatrist Dr Alan Gelenberg administered 100 mg of *tyrosine* three times daily orally for two weeks. He noted that 'The patients showed striking *improvement in mood, self-esteem, sleep, energy level, anxiety and sleep complaints.* No adverse effects were noted.' Plainly this would be an advantage to anyone with anxiety-related hypertension. A dose of tyrosine will result in either an increase in blood pressure in a person in whom it is low, or lowered blood pressure in a person in whom it is high, depending on the activity of certain nerve structures in the brain.[1]

The effectiveness of *tyrosine* in hypertensive individuals may relate to *stress reduction*. Dr Brian Morgan points out that new evidence on animals shows that under extreme stress the brain fails to produce enough norepinephrine to supply the increased demand (and remember its supply declines with age anyway). If animals thus affected were supplied with a high-tyrosine diet, brain levels of norepinephrine (formally known as noradrenaline) rose to adequate levels. *It takes between 50 and 100 mg of tyrosine per kg of body weight in animals to achieve this effect, meaning that a person weighing 100 lb would need 2½ g daily of tyrosine as a stress reducing brain medication.* This would need to be spread out through the day.[2]

A study in Japan looked at the relationship between taurine and high blood pressure. The taurine content of whole blood and urine was measured in 18 normal and 79 hypertensive patients. Blood levels were found to have no significant differences, but urine showed that normal individuals excreted almost twice the level of taurine as compared to those hypertensives who had what is known as 'essential hypertension', where blood pressure is constantly elevated, perhaps as a result of hardening of the arteries. Those hypertensives with what is known as labile hypertension, which is when there is a rapid rise and fall in blood pressure, excreted roughly the same levels as normal patients. Care was taken to exclude the possibility of any influence of kidney damage in any of the patients. The conclusion was that in high blood pressure the decrease in urinary taurine related to a depression of its formation, rather than to retention.

The results of this study suggested that a deficiency of taurine plays an important role not only in essential hypertension but in the development of atherosclerosis.[3]

Cystine and methionine both produce taurine, and animal studies indicate that liver dysfunction might have prevented adequate taurine production from cysteine in hypertensive conditions. Other possible influences of taurine on blood pressure may relate to its known effect on heart function, and also on the part it plays in the formation of bile salts, as well as its effect on platelet adhesiveness, discussed in the section on Migraine Headaches (page 137). If taurine is in short supply an increase is noted in certain blood fats (high density lipoproteins) which are harmful to cardiovascular health. Thus there are a number of different ways in which taurine can help high blood pressure and supplementation might be helpful.

References

1. Philpott, W. *Selective amino acid deficiencies*, Information pamphlet issued by Klaire Laboratories, California.
2. Morgan, B. and R. *Brain Food*, Michael Joseph, London, 1986.
3. Kohasi, N. *Urinary taurine in essential hypertension*, Japanese Heart Journal, 24(1):91–101, 1983.

Immune function enhancement, with notes on post viral fatigue and AIDS

There can be few people, in an age when we are all aware of the AIDS crisis, who are unaware of the nature and role of the immune system, the interconnecting and supremely important defence mechanisms which allow us to survive in the face of multiple threats and challenges. Diet plays a fundamental part in the preservation, enhancement and/or restoration of adequate or optimal immune function. Research has clearly shown the ability of nutritional manipulation to affect, modulate and correct the body's immune response.[1]

Space does not permit a review of the many nutritional aids which are available to boost or enhance immune function, since most of the 40 odd vitamins, minerals and trace elements have some effect, and discussion, or even listing, of these would be a lengthy exercise. The amino acids with specific immune enhancing capabilities are discussed below.

For a deeper understanding of the many elements involved in maintaining adequate immune function see Dr Michael Weiner's book *Maximum Immunity*.

Laboratory studies have shown that carnitine can be used to enhance the response of lymphocytes, the cells which play a major part in the immune system, in both humans and animals.

These were challenged by cancer inducing processes and yet when carnitine was added even in small concentrations, this was shown to be remarkably protective and beneficial.[2]

Deficiencies of amino acids in general, and of specific ones such as taurine and glutamine, are known to weaken the immune response.[3,4,5]

A number of amino acids stimulate production by the pituitary gland of a substance known as growth hormone. This has many roles to play in the processes of growth and repair, and also in immune function stimulation. *Arginine (5 to 10 g) and/or ornithine (2½ to 5 g) both stimulate production of growth hormone and can be used to enhance immune function when needed.* These should be taken either individually or together, on an empty stomach at bedtime.

CAUTION: The supplemental use of arginine by schizophrenics is undesirable.[6]

AIDS

Since AIDS is obviously the ultimate example of a compromised immune system, the uses of amino acids in strengthening the immune system generally are a guide to their value in other degenerative conditions.

Here, amino acids are of most value in general health enhancement rather than to obtain specific pharmacological effects (see numerous examples in Chapter 6), although some specific results are possible, as mentioned above. One of the major symptoms of AIDS is reduction of the body's ability to manufacture adequate protein, accompanied by steady weight loss. As we know, amino acids are the building blocks of protein.

Protein may be inadequately processed by the digestive system for a number of reasons. Thus the food consumed may fail to be reduced to its basic components suitable for absorption and utilization by the body, and if this happens these constituents should be provided in a form which is readily absorbable and usable.

Apart from amino acids with specific therapeutic roles, general free form amino acids (in other words individual, not bound together to make a particular protein) are required. ARC (AIDS related complex) and AIDS patients need this steady supply of amino acids, as almost all of them have impaired digestion and bowel function. The same applies to many chronic diseases, especially those related to what are known as auto-immune diseases (rheumatoid arthritis, lupus) and chronic bowel conditions (colitis etc).

ME

Another group of people has recently been acknowledged as being 'really ill' by medical science, after years of being considered neurotic or hypochondriac. These are the hundreds of thousands of individuals suffering from what is variously known as Post Viral Fatigue Syndrome, Myalgic Encephalomyelitis (ME), Royal Free Disease, Icelandic Disease or 'Closet AIDS', whose major symptom is an overwhelming and profound fatigue after any physical or mental effort.

The condition seems to relate to the ongoing presence of a viral agent (often Epstein-Barr or Cytomegalovirus) which results in a sort of shadow image of AIDS without its more serious, often fatal, implications. What ME and AIDS have in common is a compromised immune system, bowel dysfunction and usually (almost always in fact) Candida overgrowth.

The sort of nutrition desirable for people with ME is the same as that recommended for people with AIDS, and this is outlined below. A variety of nutritional aids as well as herbal and other remedies can help restore ME patients to normal over a period, and the very least that might be done during the recovery stage, apart from the more comprehensive approach described later in this section, would be to take free form amino acids on a regular basis. The dosage should be dependent upon the degree of exhaustion and other factors such as digestive and bowel competence, as well as continuing infections, and can be gradually reduced once energy returns. When the bowel has been damaged or is infested with protozoa and other parasites as well as yeasts such as Candida, as it almost always is in people with ME and AIDS, its permeability becomes impaired, allowing absorption of large molecules of partially digested foods which would not happen in normal digestion. This can result in a variety of toxic and allergic reactions.

Amino acid profile

A specialized analysis technique is now available in which various body fluids such as the serum of the blood or urine can be profiled to show the levels of the many amino acids in them. This knowledge is currently backed up by sufficient research to make it possible to compare the findings in any given sample with a standard 'normal' range, thus showing which of the 40 odd amino acids and their metabolites are in excess or deficit, or indeed are within normal ranges. A few companies in the UK and the US work with doctors using nutrition therapy and analyse Candida, AIDS and ARC patients, as well as those with many other serious conditions, during various stages of their illnesses. From the analysis it is possible to obtain a picture of current symptoms as well as to make a prognosis. There have been major beneficial changes in health and symptomatology as a result of the scientifically applied nutritional supplementation of amino acids which can be formulated after such profiles are completed.

The Tyson organization in the US has records of some 24 individuals with AIDS/ARC who have reversed the negative T-cell ratios and whose health has apparently been restored. At this early stage it is impossible to know their long term prognosis, although there is no reason to anticipate a decline in their regained health. However, the feeling is that it is too early to claim success in dealing with AIDS or ARC, and that such claims can only be made when there are documented records of such individuals surviving in good health for at least three years (the patient with the longest period on this programme currently has been on it for 2½ years).

A formulation of amino acids and associated nutrients suitable for a person with AIDS might include the following (exact quantities would depend upon individually assessed needs):

At least half an hour before meals:
- *A balanced amino acid formulation excluding arginine.*
- *A teaspoon of a high potency formulation acidophilus such as Superdophilus or Megadophilus in water.*

- *A teaspoon of Bifido factor (another bacterial culture which repopulates the bowel with 'friendly' bacteria).*
- *If depression is a factor, 1000 to 1500 mg of crystalline L-tyrosine together with 20 mg of vitamin B_6 (in its active form of pyridoxal-5'-phosphate) before breakfast and lunch only.*
- *400 to 500 mg of L-tryptophan together with vitamin B_6 and vitamin B_3 before lunch, evening meal and bedtime.*
  During all meals:
- *Two or three high potency formula vitamin/minerals guaranteed from yeast free sources (because of Candida) and including active forms of vitamins B_2 and B_6. These are best in capsule form as tablets may be indigestible to a person with AIDS or bowel dysfunction.*
- *One or two high potency antioxidant formulations (including Beta carotene, glutathione and corn free vitamin C).*
- *Five to ten high potency amino acid capsules (formulation varies with the needs of the individual and the presence or otherwise of Epstein-Barr virus or herpes infection – see pages 120 and 127 for why patients might need more or less of lysine and/or arginine).*
- *One or more grams of corn free vitamin C.*
- *One high potency, soy free choline supplement including in its formulation phosphatidylcholine (at least 400 mg), phosphatidylinositol (at least 200 mg), phosphatidylethanolamine (400 mg), vitamin B_6 (50 mg), methionine (50 mg), serine (50 mg) and manganese (5 mg).*
- *Lysine (500 mg) if herpes is a factor. The effect of lysine on other viruses is under investigation and it may be desirable in all viral infections.*
- *If herpes infection is not present, 500 to 1000 mg of arginine.*
  After all meals:
- *Half to one teaspoonful Lactobacillus bulgaricus.*

Three meals should be eaten daily.

If Candida is an ongoing problem then the above programme should be modified.

A nutritional programme for a person infected with HIV but not actively with ARC/AIDS conditions would include similar nutrients in highly modified doses.

These dosages are based on those used by physicians who

are treating immune deficiency conditions and Candida with apparent success and are documented in *World Without AIDS* by S. Martin and L. Chaitow.[7]

A different approach is taken by Dr Robert Erdmann. *He proposes use of a supplement which incorporates all the major nutrients known to be involved in immune functions including a number of free form amino acids. This general supplement (formula below) is taken together with supplemented vitamins A (7500 iu daily) and C (to bowel tolerance), Beta carotene (the precursor of vitamin A found in vegetables – 15 mg) and essential fatty acids (1500 mg in divided doses) derived from the plant evening primrose.*

The general amino acid supplement contains the following: *arginine 140 mg, ornithine 40 mg, glycine 40 mg, taurine 40 mg, cystine 40 mg, glutamic acid 20 mg, alanine 20 mg, tyrosine 20 mg, tryptophan 20 mg, histidine 20 mg, lysine 20 mg, as well as vitamins B_1 20 mg, B_2 20 mg, B_6 32 mg, B_{12} 200 mcg, vitamin C 40 mg, pantothenic acid 32 mg, folic acid 80 mcg, sorbitol 8 mg, magnesium 40 mg, selenomethionine 20 mcg, zinc gluconate 3.2 mg.*

Five capsules containing the above are taken half an hour before breakfast as a general 'health insurance' to maintain immune function.

This intake is increased to three times daily in cases of greater need (AIDS, ME etc).

The notes given above are meant to give a broad nutritional approach to serious degenerative disease or compromised immune function. Care should also be taken to see that the main diet consists of fresh, wholesome, largely unrefined sources of foods, with a basic low fat, high complex and low simple carbohydrate pattern. Protein intake (from lean meat, game, fish, pulses, nuts and seeds) should also be adequate. An abundance of raw food is suggested, digestion permitting. If not possible then lightly cooked (steamed or stir fried) foods are usually acceptable to a sensitive bowel.

References

1. Corman, L. *Effects of specific nutrients on immune response*, Med. Clin. North America, 69(4):759–91, 1985.

2. Simone, C. et al. *Vitamins and immunity: influence of carnitine on immune system*, Acta Vitaminologica Enzymologica, 4(1-2):135-40, 1982.
3. Sites, D. et al. *Basic and Clinical Immunology*, Lange Medical Publications, pp297-305, 1982.
4. Kafkewitz, D. *Deficiency is immunosuppressive*, American Journal of Clinical Nutrition, 37: 1025-30, 1983.
5. Masuda, M. et al. *Influences of taurine*, Japanese Journal of Pharmacology, 34(1):116-18, 1984.
6. Pearson, D., Shaw, S. *Life Extension*, NutriBooks, 1983.
7. Martin, S., Chaitow, L. *World Without AIDS*, Thorsons, 1988.

Infertility

There are a number of possible causes of infertility and amino acid therapy can only be effective in some of them. It is worth remembering that most cases of apparent infertility resolve themselves without any treatment and that evidence suggests that as large a proportion of couples attending fertility clinics achieve pregnancy as do apparently infertile couples who do not have any special attention.

In a study in Canada involving 1145 apparently infertile couples, assessed over a period of between two and seven years, pregnancy occurred in 41 per cent (597) of the couples treated by standard medical care, and in 35 per cent of the 548 couples who were untreated. It was also noted that of the pregnancies which took place amongst the treated couples, 31 per cent occurred over three months after the last medical treatment or more than 12 months after surgery. These could then be added to the 'untreated' group, in which case it could be argued that as many as 61 per cent of the pregnancies occurred independent of therapy. This should be of immense satisfaction to the many couples anxious about failure to conceive.[1]

This does not of course mean that nothing can or should be done to establish causes, which at times are simply and easily remedied. A good healthy diet supplemented with vitamin C, zinc and arginine, adequate exercise and rest, abandoning use of tobacco and severely restricting consumption of alcohol, are usually all that is required to prepare the body of the mother for conception.

Organizations which advise on preconceptual care of both parents, have statistically proven that such approaches work. This is not to say that in rare cases the heroic methods, pioneered in the UK, of conception achieved outside the womb and reimplanted, are of no value. These however can only be of value to a very limited group of people.[2]

An experimental study was carried out in which 178 men with severely deficient sperm levels and reduced motility of sperm (both major causes of male infertility) were treated with 4 g of arginine a day. Of these 111 achieved marked improvement and 21 others showed moderate improvement. The remaining 25 per cent of patients showed no improvement.[3]

Animal studies would suggest that in cases of infertility there is a deficiency of the amino acid lysine or its derivative carnitine.

Males have very high levels of carnitine in their testes, and higher levels in the bloodstream than females. When animals are deprived of lysine they become infertile due to loss of sperm motility.[4,5] Changes have also been shown (in various types of infertility) in ornithine, arginine and total amino acid concentrations in the seminal plasma.[6]

Definitive conclusions as to amino acid influence are not yet available but at the least arginine together with zinc and vitamin C should be used by those in need. See also 'Sexual Problems', page 154.

References

1. New England Journal of Medicine, 309:20, p1201f., 1983.
2. Werbach, M. *Nutritional Influences on Illness*, Thorsons, 1989; Third Line Press, California, 1987.
3. Schacter, A. et al. *Treatment of oligospermia with Arginine*, Journal of Urology, 110(3):311–13, 1973.
4. Clinical Chem. Acta, 67:207–12, 1977.
5. Journal of Nutrition, 107:1209–15, 1977.
6. Papp, G. et al. *Role of amino acids in fertility*, Int. Urol. Nephrol, 15(2):195–202, 1983.

Inflammation

There are numerous causes of inflammation which should be seen as part of the natural healing process, and should not necessarily be suppressed because of the risk of actually making matters worse. For example, there has recently been a lot of concern in medical circles over the discovery that NSAIDS (non-steroidal anti-inflammatory drugs), used over the past 20 or so years for inflamed joints, have actually worsened the problem. In addition, NSAIDS are also well known for their disastrous effects on digestive and other functions, as witness the anti-arthritic, anti-inflammatory drug Opren or the anti-rheumatic drug Butazolidin, now both withdrawn. Thus inflammation should be treated with caution and simple methods such as cold compresses or ice applications are often far safer and more effective than medication.

Sometimes however treatment is needed and here again nutrients have shown themselves to be useful, for example vitamin C, vitamin E, zinc, omega-3 and -6 fatty acids, bioflavonoids such as quercetin, proteolytic enzymes such as bromelain (from pineapple plant), alone or in combination. Certain amino acids have also been found to be useful including creatine, phenylalanine and tryptophan (D and L forms) and valine (also D and L). Both phenylalanine and tryptophan have also been shown to have powerful analgesic (pain relieving) properties, as we shall see in the secion on Pain Control (see page 143).

In animal studies it was shown that creatine was as effective an anti-inflammatory medicine as was the drug phenylbutazone in both chronic and acute situations, whilst producing no gastro-intestinal reactions. It was also noted for its pain killing (analgesic) effects.[1]

It has been suggested that D-phenylalanine is capable of reducing inflammation in much the same way as it reduces pain, because of its ability to slow down the natural breakdown of pain killing substances (endorphins and encephalins) produced by the brain and allow them to maintain their role for longer. That inflammation would be reduced in joints in much the same manner is supported by animal studies, for when such endorphins were injected into rat paws, inflammatory processes were neutralized.[2,3]

Tryptophan acts in a similar manner to phenylalanine. A study on animals found that tryptophan (L or DL) was as effective as phenylbutazone in suppressing acute or chronic inflammatory responses but not in exactly the same way. Similar studies using valine were assessed by the same researchers, with good results.[4]

For suggestions on dosage, follow the advice given in the section on Pain Control on pages 143–146.

References

1. Khanna, N. et al. *Anti-inflammatory activity in creatine*, Arch. Int. Pharmacodyn Ther., 231(2):340–50, 1978.
2. Millinger, G.S. *Neutral amino acid therapy for management of chronic pain*, Cranio 4(2):156–63, 1986.
3. Ferreira, S. et al. *Prostaglandin hyperalgesia: The opioid antagonists*, Prostaglandins, 18:181–200, 1979.
4. Madan, B. et al. *Anti-inflammatory activity of L and DL tryptophan*, Indian Journal of Medical Research, 68:708–13, 1978 and *Anti-inflammatory activity of DL-valine*, Indian Journal of Experimental Biology 16:834–6, 1978.

Insomnia

One of the major successes of amino acid therapy has been related to sleep disturbances. Drs Goldberg and Kaufman, in their book *Natural Sleep*, describe the research on tryptophan and sleep carried out by Dr Ernest Hartman of Boston State Hospital. In a series of 11 experiments conducted over a period of seven years up to 1978 Dr Hartman documented the effects of tryptophan on both animal and human subjects. All the studies compared tryptophan, at various doses, with placebo substances under double blind conditions. Tryptophan and the placebo were administered in 1 g dosage 20 minutes before bedtime, and a variety of physiological variables were recorded including changes in brainwave pattern. Those tested included both normal sleepers and insomniacs.

The findings showed a speeding up, by 50 per cent, of what is called sleep latency (the time it takes to go to sleep), and a deeper, more refreshing sleep experienced as a rule.

No side effects were noted.[1]

Hartman's report on tryptophan showed the following:

1. Increased sleep latency.
2. Doses lower than 1 g were not effective and larger doses did not make for more improvement in sleep patterns. 1 g is the dose recommended for most insomniacs although those with severe sleep problems may benefit from higher intakes.
3. Continued use usually helped those not initially showing signs of benefit.
4. If doses were kept below 15 g daily, no side effects were noted of any consequence.[2]

Research into how tryptophan works involved the administration of between 1 and 4 g before bedtime in ten male patients, all chronic insomniacs, aged between 30 and 72. There was a 30 per cent sustained relief from insomnia, with no side effects, in 90 per cent of the patients. It is thought that the formation of the calming neurotransmitter serotonin (from tryptophan) is its means of action.[3]

A double blind, cross-over study, involving 20 males, was carried out to assess the effects of a single dose of tryptophan or tyrosine, against a matched placebo. Various tests of mood state were performed. The results showed that tryptophan increased the feelings of tiredness and decreased feelings of vigour and alertness, although this was mostly imagined by the subjects with no objective evidence. Tyrosine, by contrast, was found to increase alertness and reaction times. The conclusions were that tryptophan had significant sedative properties and that, despite the feelings it created, it did not, unlike drugs, impair performance of normal tasks such as driving a car.[4]

Goldberg and Kaufman remind us that vitamin B_6 is essential for the body to use tryptophan efficiently. Dosage of 50 to 150 mg is suggested.

Tryptophan is more efficiently utilized if taken with a small amount of carbohydrate (half a biscuit for example) 20 minutes before retiring. Avoidance of caffeine, low intake of alcohol and adequate exercise and relaxation are also desirable for better sleep.

A nutritional 'sleeping tablet' combining tryptophan, vitamin B6 and two other factors, calcium and magnesium, is marketed as *Somnamin* by Larkhall Laboratories.

References

1. Goldberg, P., Kaufman, D. *Natural Sleep*, Rodale Press, 1978.
2. Hartman, L. *Report to an American Medical Association symposium*, reported in Clinical Psychiatry News, March 1985.
3. Fitten, L. et al. *Tryptophan as hypnotic in special patients*, Journal of the American Geriatric Society, 33:294–7, 1985.
4. Leiberman, H. et al. *Effects of dietary neurotransmitters on human behaviour*, American Journal of Clinical Nutrition, 42(2):36–70.

Menopausal problems

There are many different symptoms experienced by women at the change of life. The more obvious and common ones, though, are depression (see page 108), fatigue (see page 117), and hot flashes. Hot flashes and sweats can be helped by supplementation of bioflavonoids and vitamin E.[1] Tryptophan has also been shown to be useful in such cases since its deficiency may relate to menopausal depression. Low blood and oestrogen levels were found in women with this problem.[2]

In another study, methionine was found to be handled differently by pre- and postmenopausal women. The role of sulphur (which is part of methionine) was observed in ten premenopausal and ten postmenopausal women, all of whom were in good health, and the results compared with two groups of men of comparable ages. After overnight fasting each person tested was given 0.1 g of methionine per kg of their body weight. The blood was tested immediately before and eight hours after this 'loading' test for methionine derivatives including methionine itself, homocystine, cystine, and homocysteine-cysteine.

In the fasting state, before supplementation, premenopausal women and both groups of men showed similar levels of all substances tested for, while postmenopausal women had low values of the same substances.

They also displayed excessive levels of homocysteine after

the loading test, showing that before the menopause women can metabolize methionine and its derivatives more efficiently. This extra efficiency in younger women is thought to account, at least in part, for their lower incidence of cardiovascular disease during childbearing years.

Menopausal women should be careful in the degree of methionine supplementation they undertake.[3]

References

1. Werbach, M. *Nutritional Influences of Illness*, Thorsons, 1989; Third Line Press, California, 1987.
2. Editorial in the British Medical Journal, 1:242–3, 1976.
3. Journal of Clinical Investigation, 72(6):1971–6.

Migraine headaches

Migraine headaches have a number of causes, both emotional and physical.

Food sensitivities also play an important part. One study identified colouring and flavouring agents in food, alcohol, chocolate, coffee, tea, foods containing tyramine, as well as certain vitamins and minerals and foods contaminated with pesticides, as all causing headaches in sensitive individuals.[1]

In particular, the food substances which contain elements that lead to migraine in sensitive individuals include:

Nitrates found in cured and luncheon meats in large amounts. Thus bacon, sausages, ham, salami, hot dogs etc are possible foods to eliminate.

Monosodium glutamate found in Chinese cooking (restaurant food that is) and much canned food. Even normally non-headache subjects will develop a thumping skull when this food additive (the sticky glutinous stuff so obvious in much Chinese restaurant food) is present in large amounts.

Tyramine found in milk, cheese, especially mature varieties, chocolate, eggs, wheat, peanuts, citrus fruits, tomatoes, pork, pickled herrings, salted dried fish, sausages, chicken liver, beef, Italian broad beans (fava), sauerkraut, vanilla, yeast and yeast extracts, soy sauce, beer, ale, red wine, Sauternes, Riesling wines, champagne, sherry, port, and in fact in alcohol of all sorts.[2]

An experimental double blind study found that 93 per cent of 88 children with severe and frequent migraines recovered when placed on a special diet (known as an oligoantigenic diet) which eliminated foods that caused the problem. Once identified, the foods were used again, provoking headaches and thus proving the connection. The foods most suspect were dairy produce, eggs, chocolate, wheat and some meats (pork).[3]

It is believed that the actual mechanism causing the headache involves an abnormality in the function of blood platelets, allowing for increased concentration, reduced circulatory efficiency and the specific symptoms of migraine headaches. A study was conducted into the connection between *taurine* and platelet adhesiveness. Taurine is a sulphur containing amino acid and it was noted that during migraine attacks, taurine platelet levels were significantly greater than in the period of five to ten days following the headache. The researchers noted that more than one chemical species was involved in these changes apart from taurine, including serotonin (derived from tryptophan).

The significance of these findings is not clear although it would suggest that manipulation of taurine and tryptophan levels could possibly reduce migraine incidence.[4]

Erdmann and Jones suggest use of *DLPA* (see section on Pain Control, page 143) for treatment of migraine, as well as *tryptophan*, stating that its ability to help in this respect can 'be attributed to its serotonin-producing pathway and the dilatory effect it has on blood vessels – relieving the pressure areas which cause migraine by distributing the blood more widely.'[5]

Ironically, although deficiency of serotonin (which is a derivative of tryptophan) is a cause of migraine, so are excess levels of serotonin. It has been pointed out that headaches (along with fatigue and many other symptoms) increase in an atmosphere of positive ionization. This is the sort of atmospheric electrical change which occurs when a storm is coming up, or there is a strong prevailing wind, such as a mistral, or the sort of atmosphere found in centrally heated, air conditioned, or smoke polluted modern offices and apartments. It seems that production of serotonin is stimulated by this alteration in positive air ion content and that this can be effectively reversed by the simple use of a negative ionizer. These instruments are now readily available and are fairly inexpensive.[6,7]

The use of the herb feverfew reduces platelet aggregation, thus effectively inhibiting migraine attacks.[8]

References

1. Cephalgia, 2(2):111–24, 1982.
2. Harold Gelb, *Killing Pain Without Prescription*, Harper and Row, 1980.
3. Egger, J. et al. *Is migraine a food allergic disease? The Lancet*, pp865–9, 15 Oct 1983.
4. Dhopesh, V. et al. *Change in Platelet taurine and migraine*, Headache Journal, 22(4):165, 1982.
5. Erdmann, R., Jones, M. *The Amino Revolution*, Century, 1987.
6. Soyka, F. *The Ion Effect*, Dutton, New York, 1977.
7. Mann, J. *Secrets of Life Extension*, Harbor Publishing Co., 1980.
8. Johnson, E. et al. *Efficacy of Feverfew in treatment of migraine*, British Medical Journal, 291:569–73, 1985.

Multiple sclerosis

Supplementation and dietary manipulation have in many instances allowed a remission of MS symptoms. Those most investigated include the essential fatty acids derived from oil of evening primrose. Injections of thiamine have also been found useful, and deficiencies of calcium, vitamin D and pyridoxine may also be causes. Massive supplementation of B vitamins, vitamin C and many other nutrients including the amino acid phenylalanine have been reported as helpful, as has removal of dental amalgams which may be producing mercury toxicity (these should be replaced with composite dental materials including ceramic type substances). A high fat diet interferes with the production of derivatives of essential fatty acids and therefore a low fat diet is considered desirable.

The amino acid phenylalanine has been reported to be useful.[1]

In an experimental double blind trial 50 MS patients were treated with D-phenylalanine and electrical stimulation (TENS). Forty-nine of the 50 showed improvements such as better bladder control, greater mobility and less depression.[2]

Use of amino acids such as cysteine and glutathione (along with zinc, calcium and vitamin C) may assist in elimination of the toxic heavy metal mercury, which produces symptoms similar to MS, from the body.

References

1. Werbach, M. *Nutritional Influences on Illness*, Thorsons, 1989; Third Line Press, California, 1987.
2. Winter, A. *New Treatment for MS*, Neurological & Orthopaedic Journal of Medicine & Surgery, 5(1), April 1984.

Motor Neurone Disease (see also Amyotrophic Lateral Sclerosis)

One of the same factors which seems to negatively influence Amyotrophic Lateral Sclerosis is thought by researchers in the United Kingdom to be a major aggravating factor in motor neurone disease (MND), a disease which kills one person in 500, in which nerve cell degeneration results in muscle wasting.

This relates directly to an excessive build-up of the amino acid glutamate as a result, it is thought, of a deficiency of glutamate dehydrogenase (GDH), the enzyme which breaks down levels of glutamate when these become excessive. As glutamate levels rise, motor neurone cells – vital for the function of skeletal muscles – become overactive and quite literally degenerate and die through exhaustion, with the disastrous consequence of muscle wastage.

A major clinical study into this connection is being mounted, co-ordinated by the Academic Unit of Neuroscience, Charing Cross Hospital, London, involving 760 patients with MND and running for three years. This will examine the enzyme/glutamate/MND link and will treat the patients on the study using either a dummy placebo or an amino acid cocktail which will include the branched chain group valine, isoleucine and leucine which are thought to reactivate the vital enzyme GDH.

All the patients will have their muscle strength and general functional ability monitored on a three monthly basis throughout this important study.

Reference

Hunt, L. *Acids May Give Clue to Disease*, 'The Independent', 2 June 1990, p9.

Overweight problems

In the main, weight problems are caused by the problem of eating too much of the wrong food and getting inadequate exercise. Supplementation with vitamin C, evening primrose oil and coenzyme Q_{10} can assist in recovery but cannot be a substitute for correct eating and exercise. The diet should reduce, or avoid, refined carbohydrates (especially sugar) and replace these with high fibre foods and complex carbohydrates, with much of the food eaten raw. Calorie control is one aspect of reducing weight and this calls for a steady application of basic rules rather than sudden heroic efforts.

Brilliant research has shown that there are a number of ways in which amino acids can be used to help control appetite and select desirable foods at mealtime. *It may be possible to reduce a desire for sugary foods by supplementation with glutamine, which is known to be a good way of reducing alcohol craving. Dosage is between 200 mg and 1 g three times daily.*[1,2,3]

In one study, healthy young male volunteers of normal body weight were given a standard breakfast. They were then given either a capsule of *tryptophan* or a placebo 45 minutes before a buffet style (self service) lunch. Those who had taken the tryptophan selected food of significantly fewer calories than those who had taken the placebo. In another similar study consisting of 15 people the same *more desirable selection of food* was noted when 2 g or 3 g of tryptophan were given, but not when only 1 g was supplemented.

The most noticeable difference was that those receiving adequate tryptophan chose fewer bread rolls and biscuits and reported that they were not as hungry as usual.[4]

Tryptophan was also the subject of a study of 62 overweight Swiss patients, when it was used in combination with what is known as a protein-sparing modified fast (PSMF). These people were questioned as to the times when they usually experienced cravings for carbohydrates. They were asked to drink a liquid which contained either tryptophan or a placebo twice daily for three months, at times of 30 or 60 minutes before the expected craving, or at least 60 minutes after a meal. The trial was double blind in that neither the doctors nor the patients knew who was consuming tryptophan. At the start of the study all the patients were at least 40 per cent over their normal body weight. Those receiving tryptophan increased their weight loss by 3.4 kg in the first month and 2.6 kg in each of the next two months.

The best results were noted in those who were not grossly overweight at the outset.[5]

Another study was carried out on people of normal weight, and people who were slightly and very overweight, in which a combination of amino acids (*phenylalanine 3 g, valine 2 g, methionine 2 g, tryptophan 1 g*) or a placebo was given half an hour before a meal. Between 8 and 32 g of the amino acid mixture was given. *A significantly reduced intake of food was noted* in those who were above their body weight, but not in those of normal weight. This may have been the result of stimulation of the gastrointestinal hormone cholecystokinin, which is thought to have the effect of reporting to the brain that enough has been eaten to satisfy needs.

Previous studies have shown that phenylalanine has the effect of stimulating production of cholecystokinin.

It is also known that tryptophan has the same effect as a carbohydrate meal, causing the release of the neuro-transmitter serotonin in the brain, which could explain the lower selection of carbohydrate foods in overweight people who have had tryptophan supplemented.[6]

Animals allowed to choose between carbohydrate and protein rich foods not only regulate the amount of calories consumed, but also the proportion of protein and carbohydrate. Many overweight people, on the other hand, consume half their daily intake of calories in the form of carbohydrate-rich snacks, often associated with a strong craving. This may be the result of an abnormality in the brain's serotonin release process, which tryptophan can moderate.[7]

The strategy to adopt would appear to be to eat a small amount of carbohydrate together with a gram or two of tryptophan (but not if you are pregnant) about 20 minutes before a meal. This causes release of serotonin and ensures that less carbohydrate and more protein will be eaten, and that the food chosen will be more appropriate to weight loss.

For overall lowering of appetite take 700 to 1100 mg of phenylalanine to induce release of cholycystokinin within half an hour. Glutamine supplementation as discussed above will reduce sugar craving.[8]

References

1. Passwater, R. *Glutamine: the surprising brain fuel*, Educational pamphlet.
2. Goodwin, F. National Institute for Mental Health, quoted in APA Psychiatric News, 5 Dec. 1986.
3. Williams, R. *Nutrition Against Disease*, Bantam Books, 1981.
4. Heboticky, N. *Effects of L tryptophan on short term food intake*, Nutritional Research, 5(6):595–607, 1985.
5. Heraif, E. et al. International Journal of Eating Disorders, 4(3):281–92.
6. Butler, R. et al. American Journal of Clinical Nutrition, 34(10):2045, 1982.
7. Wurtman, R. *Behavioural effects of nutrients*, *The Lancet*, p1145, 21 May 1983.
8. Chaitow, L. *The New Slimming and Health Workbook*, Thorsons, 1989.

Pain control

Phenylalanine and tryptophan are the two key elements in pain control (as they are in weight control and depression). Pain should be thought of as a clear warning from the body that all is not well. Thus simply obliterating pain, whether with medication or other means of pain control such as acupuncture or electrical stimulation (TENS), without consideration as to what the alarm indicates, is poor medicine.

However, once the cause is attended to it is clearly desirable to control or reduce the level of pain. Acupuncture and acupressure techniques achieve much of their effectiveness, it is thought, via stimulation of the release of natural self-produced pain killers. These are known variously as endorphins and encephalins. It is thought that phenylalanine enhances pain relief by slowing down the breakdown of these body-produced pain relievers, giving them a longer time to act. The psychological element is also very important in pain relief: the more anxious we are, the greater we perceive the pain; the more relaxed we are, the less we perceive the pain. The same degree of pain will at different times be perceived as greater or lesser according to whether we are tense and anxious or calm and relaxed. Tryptophan and phenylalanine have both been shown to have relaxing and calming actions (see Depression, page 108). Thus these versatile fractions of protein have a dual purpose: in calming and in pain control.

An experimental study was conducted in which 43 patients mainly with *osteoarthritis*, were given 250 mg *D-Phenylalanine (DPA)*, three to four times daily, for between four and five weeks. During the last two weeks, *significant pain relief* was noted, especially in those patients with osteoarthritis. The DPA became more effective as pain decreased.[1]

In an experimental double blind cross-over study it was found that after two weeks of taking 250 mg of *DPA* three times a day, seven out of 21 chronic pain patients were able to *stop all other medication* and to note a *50 per cent reduction in pain levels*. Of those patients on a placebo only one improved while 13 patients showed no significant improvement from either DPA or placebo.[2]

Studies showed that in animal experiments previously unsuccessful acupuncture became more effective once *phenylalanine* had been supplemented, and that the effects of *pain relief* were more persistent. Similar results were obtained in humans. Administration of *phenylalanine* (2000 mg) one hour before a dental procedure resulted in an *increased threshold of pain*, which further rose after an acupuncture treatment. Pre-administration of DPA enhanced the analgesic (pain relieving) effects of humans undergoing acupuncture for low back and dental surgery, where acupuncture alone was ineffective.[3,4]

A study was conducted involving 14 cancer patients to see the effects of injection of 3 mg of endorphins (the natural pain killers of the body which DPA protects from degradation, which cannot be taken by mouth as they are destroyed in the stomach). All the patients had chronic, sleep preventing pain which has resisted powerful narcotic medication, yet after this treatment they all reported profound and longlasting *complete* relief from their pain which took about five minutes to appear and which lasted for a day and a half. There were no side effects (apart from some of the patients becoming extremely happy!).

A similar study involved women who were about to deliver their babies. They were given just 1 mg of beta-endorphin by injection and all labour pains disappeared completely, with no side effects, apart from mild drowsiness. The relief of all pain was noted for between 12 and 32 hours. It is considered that many chronic pain sufferers and people with low pain thresholds have a low level of natural endorphins. In some studies the levels have been nearly 90 per cent lower than those in people with normal tolerance to pain. DLPA retards the breakdown of endorphins and thus ensures longer natural pain relief.[5,6]

Tryptophan, in doses of 2 to 4 g daily, taken with sugar and with no protein for 90 minutes before and after taking, had *pain relieving effects*. Thirty chronic patients were involved in a double blind study in which they were placed on a high carbohydrate, low fat, low protein diet and were randomly assigned to receive 3 g of tryptophan or a placebo.

After four weeks pain thresholds had risen in the tryptophan group but not the placebo group.[7]

In an experimental double blind study involving 30 normal individuals, some received 2 g of *tryptophan* daily in divided doses and others a placebo. After eight days, dental-pulp stimulation was performed to assess levels of pain tolerance. This was much higher (that is *higher tolerance and therefore less pain noted)* in those receiving tryptophan. There were side effects of itching, nausea, weight loss and mood elevation in the tryptophan group.[8]

The form most generally available of phenylalanine is a combination of the D and L forms. This is because of the great expense of the D form alone.

The following dosage of DLPA is suggested:
Two tablets of 375 mg 20 minutes before meals three times daily. If after three weeks there has not been a considerable relief of chronic pain then the dose should be doubled. If there is still no relief then DLPA should be abandoned. There is a small percentage of failure (five to 15 per cent). Relief is usually noted within seven days at which time supplementation should be stopped until pain returns.

References

1. Balagot, R. Advances in Pain Research and Therapy, 5:289–92, 1983.
2. Budd, K. *Use of DPA and enkephalinase inhibitor in treatment of intractable pain*, Advances in Pain Research and Therapy, 5:305–8, 1983.
3. Takishige, M. Advances in Pain Research and Therapy, 563–8, 1983.
4. Mesayoshi, H. *Analgesic effect induced by phenylalanine during acupuncture analgesia in humans*, Advances in Pain Research and Therapy, 577–82, 1983.
5. Oyama, T. et al. *Intrathecal use of beta-endorphin as a powerful analgesic in man*, Advances in Pharmacology and Therapeutics, 11:39–43, 1981.
6. Fox, A. and B. *DLPA*, Thorsons, 1987.
7. Seltzer, S. et al. *Effects of dietary tryptophan in chronic maxillofacial pain*, Journal of Psychiatric Research, 17:181–6, 1982–3.
8. Seltzer, S. *Alternation of Human Pain Threshold*, Pain, 13(4):385–93, 1982.

Parkinson's disease

This condition is characterized by inability to control movement, leading to tremors, loss of use of the hands and arms, and in the worst cases, of speech. Treatment using L-Dopa is currently the favoured approach. This is not effective in all cases and in any case becomes less powerful in time. A *low protein diet* during the day increases sensitivity to L-Dopa, allowing it to be more effective in lower dosages.

Methionine has been shown to improve symptoms of

patients already deriving benefits from standard medical care. In one study, 15 patients received 1 g a day of methionine, which was gradually increased to 5 g a day, without interruption of normal medication.

After two months, ten patients showed improvements with regard to activity level, ease of movement, mood, sleep, attention span, muscular strength, concentration, and speech, and walking became less difficult. Symptoms such as drooling and trembling did not improve in all cases. There were two cases in which side-effects of either diarrhoea or nausea were noted.[1]

Another study examined the use of *phenylalanine (DPA)*. Fifteen patients received 250 mg twice daily. After four weeks neurological examination revealed significant improvements in the degree of rigidity, walking disabilities, speech problems and depression. There was no improvement in the tremor.[2]

Patients receiving L-Dopa may become deficient in tryptophan, because of competition for uptake between these two substances, and certain symptoms such as depression may appear as a result.[3]

One study, commenting on the low level of tryptophan in Parkinson patients, showed a considerable improvement in mental symptoms, when intravenous and oral supplementation was introduced. This suggests that a precautionary addition of tryptophan or protein should be given to people taking L-Dopa.[4]

In another study, 40 patients received either L-Dopa and tryptophan or L-Dopa and a placebo. The L-Dopa produced benefits in terms of reduced tremor, rigidity, walking difficulties, posture etc in both groups while a significant improvement was seen in the ability to perform certain tasks in the group also receiving tryptophan. Only this group showed an improvement in their mood and drive.[5]

It is not suggested that anyone with Parkinson's disease should experiment with amino acids on their own but that they might draw the attention of their medical advisers to the possibility of enhancing their current medication in this manner.

References

1. Smythies, J. *Treatment of Parkinson's disease with methionine*, Southern Medical Journal, 77(12):1577, 1984.
2. Heller, B. et al. *Therapeutic action of D-phenylalanine in Parkinson's disease*, Arzneim-Forsch, 26:577–9, 1976.
3. *Levodopa and depression in Parkinsonism*, The Lancet, 1:140, 1971.
4. Lehmann, J. *Tryptophan malabsorption in levodopa treated patients*, Acta Medica Scandinavia, 194:181–9, 1973.
5. Coppen, A. et al. *Levodopa and L-tryptophan therapy in Parkinsonism*, The Lancet, 1:654–7, 1972.

Peptic ulcer

Surgery is often used to deal with peptic ulcers which fail to heal under normal medical care. A number of lifestyle and dietary strategies may assist, including avoidance of smoking, milk, alcohol, sugary and spicy foods. Nutrients such as bioflavinoids, vitamin B_6, vitamin A, E and C, as well as zinc and the amino acid glutamine, may also help.[1]

An experimental double blind trial on peptic ulcer patients involved standard treatment and either glutamine (400 mg four times daily one hour before meals and before retiring) or lactose. After four weeks all seven of the patients receiving glutamine had healed, compared with seven out of 14 receiving lactose.[2]

References

1. Werbach, M. *Nutritional Influences on Illness*, Thorsons, 1989; Third Line Press, California, 1987.
2. Shive, W. et al. *Glutamine in treatment of peptic ulcer*, Texas State Journal of Medicine, pp840–3, November 1957.

Pre-menstrual syndrome (PMS)

The work of Dr G Abraham in California has identified four basic patterns of PMS (or premenstrual tension, PMT).[1] These are:

1. PMT (A) which has as its main features anxiety (hence the (A) in the title) and also mood swings, irritability, insomnia and nervous tension.

2. PMT (D) which has depression as its main feature (D) together with forgetfulness, crying and confusion.
3. PMT (C) with craving (C) for sweet foods as its main symptom, along with headache, increased appetite, palpitations, dizziness and extreme tiredness.
4. PMT (H) which relates to retention of fluid (H for hyperhydration) involving rapid weight gain (over 3 lbs) before the menstrual cycle, swelling of extremities, tender breasts and abdominal bloating.

Various nutrients can assist in all these types of PMS: magnesium, evening primrose oil, vitamin B_6, and vitamin E. Reduction of intake of salt and sugar, dairy foods and calcium, and avoidance of caffeine, nicotine and alcohol also help.

Arnold Fox advises taking amino acids in combination with other nutrients. He suggests taking 500 mg of tryptophan after breakfast and dinner, if necessary increasing this to six capsules daily, three in the morning and three in the evening (remember that this is suggested as a part of a comprehensive programme, together with vitamins B, C, E, magnesium, zinc, evening primrose oil etc at various times of the day).

He further suggests that DLPA (D and L phenylalanine) be taken as follows. If weight is more than 110 lbs, he suggests 375 mg with breakfast and lunch. If no result is forthcoming he increases this to include 375 mg with dinner. This is done for four to seven days before PMS symptoms usually appear in the cycle, stopping at the end of the menstrual flow.[2]

References

1. Abraham, G. *Premenstrual tension problems,* Obstetrics and Gynaecology, 3(12):1-39, 1980.
2. Fox, A. and B. *DLPA,* Long Shadow Books, 1985.

Protection of cells from toxicity and radiation

Over the past few years there has been increasing evidence as to the ways in which certain nutrients can offer protection to cells when these are exposed to what could be dangerous levels of toxic materials or radiation.

Dr I Brekhman reports from the Soviet Union that as part of the country's space programme over 25,000 different chemical substances have been analysed and examined in order to discover effective protective substances against the effects of radiation. The two amino acids which are now incorporated into the 'cocktail' of nutrients given to cosmonauts are histidine (1 to 2 g per day) and tryptophan (for relaxation, not radiation protection).[1]

Radiation damage occurs in large part due to release of what are called free radicals. These electrically charged entities are capable of causing cellular damage on a large scale. The normal body defence against these includes glutathione peroxidase, which is itself dependent upon selenium and the amino acid cysteine (as well as glutamic acid and glycine). Cysteine is itself a powerful antioxidant (destroyer of dangerous free radicals) and when combined with vitamin C and B_1 it has a protective effect on cells exposed to radiation.

During periods of exposure to radiation of any sort a dosage of 1 to 3 g daily of cysteine, and 1 or 2 g of glutathione, taken away from mealtimes in divided doses with diluted fruit juice, is suggested.[2]

NOTE: Diabetics should not take cysteine without guidance, and manic depressives should not take histidine.

In an attempt to assess the protective function of a variety of nutrients, human cells (lymphoblasts) were cultured and then exposed to toxic substances. The presence on their own of the amino acid taurine and the mineral zinc protected cells from retinol-induced damage, and when these nutrients were used together they abolished the swelling and enhanced cell function by 55 per cent. Vitamin E was also effective in protecting cells in this way. When the three substances were used together, they afforded *complete* protection.

It should be noted that we lose taurine under certain conditions, including myocardial infarction, skeletal damage, physical and emotional stress, high alcohol consumption, and deficiency in zinc.[3]

One of the most toxic compounds known is carbon tetrachloride which can cause massive damage to the liver. The amino acids aspartic acid, methionine and tyrosine were found to protect against liver damage in animals when given 30 minutes before exposure to the toxin. Aspartic acid, tyrosine and cystine also offered protection when given up to six hours after exposure. The toxin was still present in the liver but the tissues were protected.[4]

Glutathione and cystine protect tissues against the ill effects of tobacco smoke and of alcohol; however, they should not be regarded as antidotes to smoking and excessive drinking.

References

1. Brekhman, I. *Man and Biologically Active Substances*, Pergamon Press, 1980.
2. Saksonov, P. Antiradiation Protection (in Foundations of Space Biology and Medicine), pp317-47, Nauka Moscow, 1975.
3. Pasantes, M. et al. *Protective effect of taurine, zinc and tocopherol on retinol induced damage to human lymphoblasts*, Journal of Nutrition, 114(12):2256-61.
4. British Journal of Experimental Pathology, 64(2):166-71, 1983.

Restless leg syndrome

This condition may relate to folic acid deficiency and is often improved if folic acid and vitamin E are supplemented and caffeine containing substances are avoided.

The condition is often accompanied by (or accompanies) depression and other health problems, and can be of a severe nature, not allowing the sufferer any peace, with a constant need for the leg to 'jump' and to move around. Symptoms are often worst at night.

Tryptophan has been found useful in treating cases of restless leg syndrome. One elderly gentleman, who suffered from very high blood pressure and kidney failure (he had for several years been having dialysis), had also been having for the past three years symptoms of crawling, tingling and burning sensations in his legs, especially in the evenings and when resting during the day, resulting in insomnia and depression. No neurological causes were found for the symptoms and various drugs proved ineffective. One gram of L-tryptophan twice daily cleared both the symptoms affecting the legs and the insomnia within three days, with no side effects.

In another case an elderly man with chronic obstructive lung disease, who was receiving steroid (cortisone) type medication, had a two-year history of restless leg syndrome, with resultant depression and insomnia. Medication was ineffective, but when he was given 2 g of L-tryptophan at night this produced dramatic relief of both the insomnia and the restless leg symptoms, within four nights, with no side effects.[1]

References

1. Sandyk, R. L-tryptophan in treatment of restless leg syndrome. Letter to the American Journal of Psychiatry, 143(4):554–5, 1986.

Rheumatoid arthritis

This is a severe and often crippling disease which is not fully understood. In many instances nutritional manipulation may help, including a diet low in fats and one that excludes food from animal sources (a vegan diet). Care should be taken to ensure optimal levels of nutrient intake.

Several nutrients have been found helpful including vitamin C, vitamin B_5 (pantothenic acid), selenium, zinc, and sulphur. Particular food sensitivities may be a factor in rheumatoid arthritis and supplementation of omega 3 and 6 fatty acids as well as various nutrients such as bromelain and quercetin have been reported to give symptomatic relief.

Amino acids are useful in a number of ways. For instance, it has been found that levels of histidine are very low in rheumatoid arthritis patients although other amino acids are in normal supply.[1,2] Supplementation with histidine has been found helpful, although not dramatically so. In an experimental double blind study patients were treated with either 4½ g histidine daily or a placebo. After 30 weeks there was a biochemical change showing benefits for those patients receiving histidine and there was some evidence of improvement in those patients with a long duration of the disease.[3]

Phenylalanine (see dosages in Pain Control section, pages 143–148) is reported to have dramatically reduced symptoms in a 47-year-old female patient with rheumatoid arthritis.[4]

Tryptophan has been effective in reducing swelling in animals with rheumatoid arthritis.[5]

References

1. Journal of Clinical Investigation, 55:1164, 1977.
2. Journal of Chronic Disease, 30:115, 1977.
3. Pinals, R. et al. *Treatment of RA with L-histidine*, Journal of Rheumatology, 4(4):414–19, 1977.
4. Anaboli·m, 4(2), 1985.
5. Journal of the American Podiatry Association, 70(2):65, 1980.

Schizophrenia

This major mental illness has been found to have at least two distinct forms, one with a very high degree of histamine in the brain and one with a very low level. These two groups together account for fully two thirds of all schizophrenic individuals.

The high histamine schizophrenic is called histadelic. These are usually suicidally depressed. Methionine may be used to decrease levels of histamine (see section on Sexual Problems, page 154) since it detoxifies the body of excess levels. In addition to methionine, Pfeiffer and Iliev, the main researchers into this application of amino acid therapy, suggest using magnesium, zinc and calcium lactate.

Those schizophrenics with very low levels of histamine are known as histapenics, and supplementation of histidine can result in balancing this deficiency. Pfeiffer believes that histamine is a neurotransmitter for some as yet unidentified region of the brain.

As noted in the section on Behavioural Modification (page 102), aggressive behaviour in schizophrenics was modified by supplementation of tryptophan in a double blind cross-over study.[1]

Naturally in a condition as serious as schizophrenia such supplementation should be directed by an expert who can monitor the progress of the individual. This is not a suitable condition for self-medication.[2]

References

1. Morand, C. et al. Biological Psychiatry, 18:575–8, 1983.
2. Pfeiffer, C. *Mental and Elemental Nutrients*, Keats Publishers, New Canaan, 1975.

Sexual problems

We have already examined the role of arginine in infertility (page 131). This section relates more to frigidity and impotence, as well as to a failure to experience orgasm and premature ejaculation.

The experience of sexual arousal involves histamine release, as does the experience of an orgasm itself. If histidine levels are low, histamine production will be reduced leading to failure of orgasm in both sexes. Histidine is therefore crucial in helping people with this problem.

Major research into this subject was conducted by Dr Carl Pfeiffer who established that *additional histidine given to women in doses of 500 mg before each meal three times daily resulted in restoration of enjoyment of sexual intercourse.* (He also found that histidine taken 4 g a day can help women regulate their periods.)

In males, excessive histidine was found to result in premature ejaculation, leading to frustration. *Where this is a problem Pfeiffer found that supplementation with 500 mg methionine, together with 500 mg magnesium and 50 mg vitamin B_6 helped to normalize the excessive levels of histidine, and therefore the problem of premature ejaculation.*[1]

References

1. Pfeiffer, C. *Mental and Elemental Nutrients*, Keats Publishers, New Canaan, 1981.

Wound healing

Among the nutrient aids to wound healing are zinc, vitamins C and E and the amino acid arginine.

In an animal study involving rats a one per cent dietary supplement of arginine improved wound healing. Weight loss after the wound was reduced, the healing process was accelerated and the thymus gland, which is important in the body's defence (immune) function, increased in size. (Arginine is a stimulator of growth hormone from the pituitary and this is considered the means whereby it enhances wound healing.) The researchers state in conclusion, 'We suggest that supplemental arginine may provide safe nutritional means to improve wound healing and thymic function in injured and stressed humans.'[1]

References

1. Barbul, A. et al. *Wound healing and thymotrophic effects of arginine*, American Journal of Clinical Nutrition, 37(5):786, 1983.

CHAPTER 8

Tryptophan: the Truth

This is a cautionary tale which will be much enjoyed by conspiracy addicts. An epidemic of a potentially fatal condition, Eosinophilia Myalgia Syndrome (EMS), has been linked in the USA and elsewhere with the taking of what appear to be contaminated tryptophan products.

The initial connection was made by the Center for Disease Control (CDC) in the USA, leading fairly rapidly to a voluntary withdrawal of all tryptophan products of any strength, from health stores and pharmacies. Even tougher marketing guidelines were then sought by the Food and Drug Administration in the USA, and the Department of Health in the UK, leading to the virtual disappearance of this nutrient for the time being, unless prescribed by a medical practitioner.

There are suggestions, from people deeply involved in the health food industry, that this situation is tailor-made for regulatory agencies to exploit in their desire to have most supplements, and all amino acids, made available by prescription only. However, the symptoms of EMS are sufficiently ghastly to warrant strong action, ranging as they do from severe muscle and joint pain, to breathing difficulties, swelling of the arms and legs, skin rashes, fatigue, cough and sometimes fever. In some instances, congestive heart failure results, as well as progressive weakness and paralysis, and even death.

Several thousand people have been affected by EMS in the US but only three to date in the UK, all using US made products based on Japanese raw materials, *all of which came from just one supplier.*

Many of those closely involved in this tragedy are convinced that the root cause of the problem is contamination, and nothing very much to do with pure tryptophan itself – which is richly present in every egg, piece of cheese, meat or fish, and a large number of vegetable products.

The story itself is a tangled one, which does little to inspire confidence in regulatory agencies such as the Food and Drug Association and Department of Health. Tryptophan has been arguably the best selling and most useful of all amino acid supplements, with its wide application in cases of depression, insomnia and weight control.

For reasons which are unclear to all but the most knowledgeable, almost all amino acid raw materials derive from vast Japanese chemical manufacturers. In April 1990 *The Washington Post* let it be known that all the tryptophan related to EMS, marketed under many different labels, derived from just one Japanese company, Showa Denko, which at that time had also been named in a $20 million lawsuit brought by an EMS patient from Oregon.

The company stated that it had stopped production of tryptophan in late 1989, after the first of four deaths from EMS were reported in the US. They also made the comment: 'It is important to note that studies thus far do not independently identify the precise cause of EMS.' Japanese government agencies are also actively examining all possible links.

Dr Jonathen Collin, editor of the widely read *Townsend Letter for Doctors*, addressed the issue in May 1990: 'The argument raised by the [supplement] industry and proponents [of tryptophan] is that it has a long and safe history of use without any reporting of an eosinophilia condition.' Collin reports that rats which had had their adrenal glands removed displayed toxicity when given tryptophan, which might alert us to a possible danger from the amino acid when supplemented to people with underactive adrenal function.

He also notes a study published in 1990 in *The American Journal of Nutrition* by Dr Donald Mathias, which reports that high doses of tryptophan caused no liver damage in the animals being studied, unless they were exposed to highly toxic substances such as carbon tetrachloride. Collin also counters the claim published in the 18 March issue of *The Journal of the American Medical Association* which asserts that a metabolic by-product of tryptophan – kynurenic acid – could be the primary causative factor of EMS. This evidence, says Collin, is purely epidemiological and runs counter to other expert opinion (he quotes Dr Sandy Marke of NIMH) which holds that kynurenic acid can have little if anything to do with EMS. Collin believes that the FDA will do its best to find a reason for banning public sale of tryptophan, and that it will try to use this as a means for controlling all vitamin manufacture and distribution.

Collin further asserts that contamination has indeed been found by official agencies and that these contained a mucopolysaccharide of the bacteria E.Coli; however, 'The contamination report has not been made available'. The final twist in the tale as told by Dr Collin is that two separate physicians have reported that they have successfully treated EMS, using among other supplements, tryptophan itself.

In April 1990 Dr Russell Jaffe reported in the *Townsend Letter* that, using 'uncontaminated tryptophan' (along with vitamin C and other nutrients), a case of EMS was successfully treated. Writing in *Advances in Therapy* (to be published in 1990) Dr Christopher Caston of South Carolina discusses 20 cases of EMS who responded well to a regime which included tryptophan, vitamin B_6 and vitamin C, as well as small doses of corticosteroids. Dr Jaffe's case is that if tryptophan is the culprit it is not feasible that further doses of the toxin would rapidly result in good health being restored.

While the research and debate continues, those suffering symptoms ranging from insomnia and depression to PMS and weight problems, cannot obtain legitimate supplies of tryptophan. This may not be a threat to their lives but certainly impinges on the quality of their lives. The simple fact is – as Dr Collin reminds us – that most proteins contain a substantial level of tryptophan, and that this is being concealed from the public: 'Otherwise the current hysteria may lead to avoidance of all meats and dairy products.'

A leading US manufacturer of high-quality amino acid

supplements, Dr Don Tyson states (personal communication to the author): 'A great deal of anger and confusion reigns over the tryptophan story. To date we have never had one case of EMS in well over 1.5 million people who have consumed our tryptophan products. We are conducting research with physicians using tryptophan (on prescription only now) with patients who have no history of EMS. Some have seen EMS with patients taking other brands [same Japanese source] but never ours.' One of the physicians who uses Tyson products is Dr Caston, who has successfully treated EMS using tryptophan. In previous studies Dr Caston, a highly respected researcher, has shown tryptophan to be capable of reversing obesity which had been induced by the use of a powerful tricyclic antidepressant drug. Writing in *Advances in Therapy* (Vol.4 No.2 March/April, 1987) Dr Caston described a case of a woman whose obesity was safely normalized by use of this essentially safe amino acid. He has also shown tryptophan's enormous potential for good when used on people suffering delirium after coronary bypass surgery (*Advances in Therapy*, Vol.6 No.4 July/August, 1989). Dr Tyson had previously informed the author of his grave disquiet as to the state of some amino acid products on the market, and of his unsuccessful efforts to alert the FDA.

In the UK Optimax and Pacitron, two anti-depressants containing tryptophan, have recently been withdrawn from use. As yet, there has been no official announcement from the US or Japan on the possible contaminants. It can only be hoped that all parties can learn from this story and that tryptophan can safely appear in the market as a food supplement once more.

CHAPTER 9

Amino Acids: Summary of Functions and Therapeutic Uses

This chapter looks in detail at those free form amino acids which can be safely used for supplementation. The ways in which these have been, and can be, utilized is described more fully in Chapter 7, which looks at individual health complaints, and in which research evidence as to the benefits derived from use of amino acids is presented, as is information about dosage.

Before using an amino acid therapeutically read about its usefulness in Chapter 7, and also refer to the notes given in the chapter about its many different characteristics. This will give you an understanding of the substance and alert you to possible problems (e.g. clashes with other medicines) and also details of how to enhance its effectiveness by combining it with other nutrients and methods (best taken with vitamin C, or with another amino acid etc).

When to take an amino acid, or a combination of them, is also a key element in its effectiveness. With very few exceptions, which are discussed in the text, this should always be away from mealtimes, otherwise amino acids will be competing for absorption with the proteins present in the food of the meal. Free form amino acids are readily and speedily absorbed when taken an hour before or an hour after a meal, with plain water.

Not all of the many functions of amino acids which go to make up, and which are at work in, the body, have as yet been clearly identified, and by no means all of the therapeutic uses have as yet been discovered. The following summary of the characteristics, functions and uses of amino acids is not exhaustive, since some amino acids have been ascribed only minor roles thus far.

Just as we have come to be aware of the nutrient content of food in terms of vitamins and minerals (oranges and lemons for vitamin C etc), so in time we shall become familiar with amino acids and their different balances found in certain foods. For instance, already those people following a high lysine/low arginine diet for conditions such as herpes are able to select their foods (no chocolate, low intake of nuts and wholegrains, high intake of fruits, fish and chicken etc) appropriately. Hopefully in time this will become second nature to those of us whose unique biochemical requirements demand that to be healthy a particular balance of amino acids is best. This will include a great many people.

Supplementation will be seen to be a logical and safe approach to achieving good health (as long as basic care is taken over factors such as their occasional undesirable combination with certain other substances, and the ways they should be used in certain conditions, all of which are listed in various places in the book).

Use this book as a guide to where these marvellous protein fractions can help you, by discovering their power and understanding their potential.

Arginine

- An essential amino acid for children but not for adults.
- Secreted by the anterior pituitary gland.
- Stimulates human growth hormone (HGH) which stimulates immune (defence) function.
- Accelerates wound healing.
- Plays major part in urea cycle which detoxifies ammonia from the system (in this cycle it is converted to ornithine and then back to arginine).
- Necessary for normal sperm count.
- Involved in glucose (sugar) control mechanisms in the blood (Glucose Tolerance Factor).
- Enhances fat metabolism.
- Involved in insulin production.

Therapeutic uses
- Benefits to arthritics.
- Inhibition of tumour development.
- Helpful in some forms of infertility.
- Stimulates production of T-cells (major part of immune system) and enhances wound and burn healing.

Deficiency
- Infertility in males.
- Premature ageing.
- Toxicity and increase in free radical activity.
- Overweight.

Excess
- Enhanced virus replication (eg herpes simplex) unless adequate lysine also present.
- Aggravation of certain forms of schizophrenia.

Genetically acquired disorder
- Hyperargininaemia, which can be satisfactorily treated medically.

Caution
- Schizophrenics should use with caution.
- Avoid with herpes.

Aspartic acid

- Has protective functions for the liver and assists in detoxification of ammonia.
- Promotes uptake of trace elements in the gut and is involved in the energy cycle.
- Acts to transport magnesium and potassium to cells and is part of the sweetener aspartame (together with phenylalanine).

Therapeutic use
- Pronounced relief of fatigue has been noted in three quarters of patients supplemented with potassium and magnesium aspartate (1 g daily) (see page 117).

Histidine

- Is metabolized into the neurotransmitter histamine, which is involved in smooth muscle function, and contraction and dilation (expansion) of blood vessels.
- Is required for sexual arousal.
- Helps maintain the myelin sheaths which insulate the nerves and is required by the auditory nerve for good function.
- Stimulates production of red and white blood cells.
- Two forms of schizophrenia have been noted, which are characterized by either excessive or low levels of histamine in the body (brain).
- Best taken with vitamin C.

Therapeutic uses
- Protects against radiation damage (it was used in the Russian space programme).
- Chelates (helps to remove) toxic metals from the body.
- Has been successfully used in the treatment of rheumatoid arthritis.
- Useful together with vitamins B_3 and B_6 in normalizing problems of poor sexual arousal.
- Effective in treating ulcers in the digestive tract.
- Effective in treating nausea during pregnancy.

Deficiency
- Leads to poor hearing or deafness.

Genetically acquired disorder
- Histidineamia, which can be satisfactorily treated medically.

Caution
- Should be used cautiously by manic depressive patients who have elevated levels of histamine (see page 153).
- Women with severe premenstrual depression should avoid histidine supplementation.

Leucine and isoleucine

- Essential amino acids.
- Together with valine, they comprise the group of amino acids known as branched chain amino acids.

- They should always be supplemented in combination with each other unless a particularly strategy is being adopted therapeutically.
- Both are commonly deficient in amino acid profiles of chronically sick individuals.
- Isoleucine is useful in formation of haemoglobin.

Therapeutic uses
- Leucine is useful in Parkinson's disease.
- D-leucine may be effective as enhancer of pain killing effects of naturally produced endorphins (as is phenylalanine).

Deficiency
- Common in cases of chronic physical and mental disease.

Excess
- May predispose to pellagra.

Valine

- An essential amino acid which is needed for normalizing the nitrogen balance in the body.
- The third (together with leucine and isoleucine) of the group known as branched chain amino acids.
- Vital for mental function, muscle coordination and neural function.

Therapeutic uses
- Helpful in cases of inflammation.

Excess
- Sensations of 'crawling skin' and hallucinations.

Deficiency
- Nervousness.
- Poor sleep patterns and mental symptoms.
- Negative nitrogen balance (toxicity).

Genetically acquired disorders
- Hypervalinaemia.
- Methylmalonic aciduria.

- MSUD (maple syrup urine disease). Presents in newborn babies with vomiting, lethargy, tight muscles, fits and unless treated correctly, death follows very shortly afterwards. Some forms begin later in life. A special diet low in the branch-chained amino acids is needed and in some forms of the disease, the vitamin thiamin (B_1) is also helpful.

Lysine

- An essential amino acid often low in vegetarian diets.
- Important for children's growth and development.
- Involved in synthesis of the amino acid carnitine (see below), therefore important in fat metabolism.
- Helps in the formation of antibodies to fight disease.

Therapeutic uses
- Has been shown to be effective in treatment of herpes simplex virus, especially when combined with vitamin C and a low arginine diet. This pattern is also thought to decrease chances of atherosclerotic changes (see page 98).
- Enhances concentration.

Deficiency
- Fatigue, dizziness, anaemia, visual disorders, nausea.
- Dietary deficiency of lysine leads to increased calcium excretion and therefore higher danger of kidney stones.

Genetically acquired disorder
- Hyperlysinaemia, which can be satisfactorily treated medically.

Phenylalanine

- An essential amino acid.
- The precursor (parent substance) of tyrosine and hence of dopamine, norepinephrine (noradrenaline) and epinephrine (adrenaline) for which vitamins B_6 and C are required in the biochemical conversion processes. These substances control or affect heart rate and output, blood pressure, oxygen consumption, blood sugar levels, fat metabolism and many functions of the brain.

- Cannot be metabolized without adequate vitamin C.
- Required by the thyroid for normal function.

Therapeutic uses
L-phenylalanine
- Stimulates production of cholescystokinin, inducing satiety (feeling of having eaten enough). It is therefore useful in weight control.
- Acts as an antidepressant.

D-phenylalanine
- Powerful non-toxic, non-addictive enhancer of endogenous (produced by the body itself) pain killers (by slowing down their normal degradation).
- Reduces symptoms of multiple sclerosis and Parkinson's disease.
- Acts as an antidepressant.
- May improve memory, concentration and mental alertness.

D-L phenylalanine
- Pain control.
- Antidepressant.
- Treats symptoms of rheumatoid arthritis.
- Treats the depigmentation in the skin condition vitiligo (together with ultra violet light).

Deficiency
- In childhood, leads to tyrosine deficiency and therefore mental retardation, as well as melanin deficiency which makes eczema more likely.
- In adults, leads to emotional disorders, weight gain, circulatory problems.

Genetically acquired disorders
- Hyperphenylalanineaemias which include phenylketonuria (PKU), a common amino acidopathy (affecting one in 14,000 babies in North America). PKU leads to mental retardation and lack of pigmentation. This is treated by diets low in phenylalanine. Rapid detection is essential if prevention of the worse symptoms is to be effective.

Dangers of phenylalanine
- Should not be used by anyone currently taking any of the monoamine oxidase inhibitor drugs (MAO).
- Should be used with caution by anyone with high blood pressure.
- Should be avoided by phenylketonurics and by pregnant or lactating women.

Tryptophan

- An essential amino acid, needed for synthesis of nicotinic acid (vitamin B_3) in the body and the precursor of the neurotransmitter serotonin which is a calming sedating substance essential for normal mood and sleep patterns.
- Influences the amount of protein chosen at meals and is therefore used to control weight reduction.
- Uptake of tryptophan by the brain is enhanced by vitamin B_6 and vitamin C.
- Acts as a mood stabilizer (calms agitation, stimulates depressed individuals).
- The less tryptophan the greater the degree of emotional disturbance.

Therapeutic uses
- Useful in some forms of vascular migraine.
- Has anti-depressant potential.
- Useful in weight control.
- A powerful sleep enhancer when combined with magnesium and vitamin B_6.
- Menopausal depressive conditions.
- Has powerful pain killing effects.
- Useful for individuals with Parkinson's disease who are taking levo-dopa drugs.
- Helps symptoms of 'restless legs syndrome'.
- May be helpful in treating rheumatoid arthritis (animal studies).
- May help patients with tardive dyskinesia.

Caution
- May be dangerous if used in pregnancy.
- May aggravate bronchial asthma.
- May aggravate the auto-immune condition lupus.

Deficiency
- Insomnia, mental disturbance and depression.
- Poor skin colouring and tone, and brittle fingernails.
- Indigestion.
- Craving for carbohydrate.

Carnitine

- This is synthesized in the liver from lysine and methionine, vitamin C being essential for its conversion.
- Men have greater physiological need for carnitine than women.
- Influences sperm motility; reduces triglyceride levels in the blood and protects the heart against myocardial infarction by removing free fatty acids.
- Has a major role in transferring fatty acids into cells where they are used as energy sources.
- Aids in mobilizing fatty deposits in obesity and helps in removal of ketones (fat waste products) from the blood.

Therapeutic uses
- Useful in treating some forms of infertility.
- Reduces high levels of triglycerides in the blood.
- Reduces surface fats in conditions such as cellulite.
- Helpful in circulatory problems such as intermittent claudication.
- Reduces feelings of fatigue and muscle weakness.
- Helpful in treatment of fatty liver degeneration and alcohol damage to liver.
- May be useful (with other nutrients) in helping glucose tolerance in diabetics.
- Also used in cardiac disease (especially myocardial ischemia – lack of oxygen reaching heart muscle); muscular dystrophy and other myopathies and neuromuscular diseases; obesity.

Dangers and side effects
- Two thirds of patients treated using carnitine report gastrointestinal side effects or increased body odour which disappeared or diminished with continued use at a lower dosage.

- Especial care should be taken in supplementing carnitine to people with kidney damage.

Tyrosine

- Derived from phenylalanine and is a precursor of the thyroid hormones, as well as of dopa, dopamine, norepinephrine (noradrenaline) and epinephrine (adrenaline).
- Aids in normal brain function and in treating abnormal brain function as a supplier of neurotransmitters.

Therapeutic uses
- Useful in Parkinson's disease and some cases of depression which are not amenable to treatment with tryptophan.
- Small doses of tyrosine are often more effective than large ones in increasing brain neurotransmitter levels.
- Effective in alleviating hay fever and grass allergies.

Glutamic acid and glutamine

- Under certain conditions, these may become essential amino acids.
- They are the dominant amino acids of the cerebro-spinal fluid and serum.
- Glutamine readily passes through the blood/brain barrier (glutamic acid does not).
- Glutamic acid is readily converted from glutamine and is a uniquely powerful 'brain' fuel.
- Glutamic acid gives rise to GABA, a calming agent in the brain and possibly a neurotransmitter.
- Glutamic acid is required for manufacture of the B vitamin folic acid.
- Glutamic acid is a component of the glucose tolerance factor.
- Glutamine is useful in maintaining the body's nitrogen balance.
- Needed for what is called transamination (the production of other non-essential amino acids).

Therapeutic uses

Glutamic acid
- Used in treating childhood behavioural problems.
- Re-converts to glutamine and thus detoxifies the brain from ammonia.
- Has been shown in laboratory studies to dissolve or retard formation of kidney stones.

Glutamine
- Protects the body from the effects of alcohol and decreases the desire for it, and in some cases for sugar.
- Helps heal peptic ulceration (400 mg 4 × daily before meals and before retiring).
- Useful in cases of depression.
- Blunts carbohydrate cravings, therefore helpful in treatment of obesity.
- May help in some cases of schizophrenia and senility.
- Treats fatigue.
- Used to raise IQs and for memory improvement.

Deficiency
- Can lead to cantankerous and grouchy behaviour.

Toxicity
- May result from excessive glutamine intake (over 2 g daily may lead to manic behaviour).

Methionine
- An essential amino acid containing sulphur.
- A powerful antioxidant preventing free radical damage to tissues.
- Assisted by vitamin B_6.
- Helps produce choline and adrenaline, lecithin and vitamin B_{12}.
- Assists gallbladder function through synthesis of bile salts.
- The precursor of the amino acids taurine, cystine and cysteine.
- Acts to detoxify heavy metals from the body and also excessive levels of histamine (which is part of histadelic schizophrenia symptomatology).

- Strengthens hair follicles.
- Detoxifies the liver, preventing buildup of excess fats.
- Essential for selenium bioavailability in the body.

Therapeutic uses
- Helps relieve arthritic and rheumatic symptoms.
- Useful in cases requiring detoxification and anti-oxidation.
- May retard cataract development.
- Helpful in some cases of Parkinson's disease (1 g daily, rising to 5 g for two months).
- Detoxifies excessive histamine levels found in some forms of schizophrenia.
- May be useful in gallbladder problems relating to oestrogen excess, resulting from contraceptive medication.

Deficiency
- Leads to poor skin tone, hair loss, toxic waste buildup, fatty infiltration of the liver, anaemia, retarded protein synthesis, atherosclerosis.

Genetically acquired disorders
- Hypermethionineaemia for which therapy is not indicated.

Taurine
- This neurotransmitter is manufactured in the body from methionine or cysteine in the liver, and vitamin B_6 is needed for its synthesis. One of the sulphur rich group of amino acids, the major supply should come from the diet, where it is only found in foods of animal origin.
- Women require more taurine than men, since female hormone oestradiol is found to inhibit its synthesis in the liver.
- Interacts with bile salts to maintain their solubility, and with cholesterol, preventing gall stones.
- Taurine levels rise in serum as zinc levels decrease, leading to low brain levels of taurine which are undesirable. Zinc is therefore vital to its use.
- The most prevalent amino acid in the heart, it helps

conserve potassium and calcium in the heart muscle thereby helping it to function better. Research continues to ascertain its precise role(s) in heart function, which is thought to be profoundly important.
- Influences insulin and blood sugar levels.
- Greater concentrations of taurine are found in the pineal and pituitary glands after exposure to natural full spectrum light. People deprived of this may become mentally impaired and depressed, which may relate to taurine lack.
- Increases in use when the individual is under stress.
- Only found in the L form.
- The second most prevalent amino acid in human milk but is poorly supplied in cow's milk.

Therapeutic uses
- Helpful in some types of epilepsy.
- Cardiac conditions such as congestive heart failure and atherosclerosis, stress and eye problems; compromised immune function.
- Also useful in gallbladder disease.
- Claimed to enhance IQ levels in Down's syndrome children (together with other nutrients).

Deficiency
- In children, may lead to epilepsy.
- In adults, when there is also zinc deficiency, may lead to eye problems.

Cysteine

- Derived from methionine or serine in the liver.
- A major sulphur containing amino acid (together with methionine and taurine).
- A powerful antioxidant.
- Part of tripeptide glutathione (see below) which is itself part of what is known as glucose tolerance factor (GTF) as well as being a major detoxifying agent.
- Should not be confused with cystine which is a similar but not identical substance which does not possess the antioxidant qualities of cystine.

- Converts to cystine in the absence of adequate vitamin C.
- Cystine and cysteine are vital for adequate use in the body of vitamin B_6.
- In chronic disease the formation of methionine into cysteine is often prevented.
- Contributes towards the strength of the hair (over 10 per cent of hair is cysteine) and also in enzyme and insulin production.
- Skin texture and flexibility are related to cysteine function by virtue of free radical inactivation.

Therapeutic uses
- Cysteine can usefully be supplemented in all cases of chronic disease.
- Removes heavy metals from the body, and protects against the effects of alcohol, cigarette smoking and pollution etc, by ensuring detoxification of acetaldehyde.
- Useful in iron deficiency, and is helpful in prevention of cataracts.

Caution
- Cysteine should be used with caution by diabetics.

Cystine

- This is part of insulin molecule.
- One of the sulphur rich amino acid group.
- Acts as a heavy metal chelator.

Therapeutic uses
- Useful in treatment of skin problems such as psoriasis and eczema.
- Helps tissues to heal after surgery.

Caution
- People predisposed to kidney or liver stones should take care in use of cystine.

Glutathione

- A tripeptide made up of the amino acids cysteine, glutamic acid and glycine.

- Inhibits damage to fat cells induced by free radical activity and retards the ageing process.
- Neutralizes dangerous atmospheric substances such as petrocarbons and chlorine.
- Because of its cysteine content it has a sulphur element which accounts for its detoxification potential.

Therapeutic uses
- Has been shown to protect the liver against alcohol induced damage.
- Protects against radiation effects.
- Chelates heavy metals from the system.
- Causes regression of tumours in animals.
- Has been noted to be in short supply in tissues of diabetics where it acts on accumulated dehydroascorbic acid to produce vitamin C.

Gamma-aminobutyric acid (GABA)

- A non-essential amino acid formed from glutamic acid.
- Helps regulate nerve function and enhances the ability of vitamin B_3 (niacinamide) to act.

Therapeutic uses
- Induces calmness and tranquillity in cases of manic behaviour, acute agitation, schizophrenia, epilepsy and high blood pressure.
- Helpful in cases of enlarged prostate gland by stimulating the release of the prolactin hormone from the pituitary, resulting in reduction in size.

Glycine

- A non-essential amino acid which is part of the tripeptide gluthathione, which is a detoxifying agent, especially for the liver.
- Essential for synthesis of bile acids and nucleic acid.
- The simplest and sweetest of amino acids and is used as a sweetener.

Therapeutic uses
- No proven individual therapeutic uses but together with other amino acids may be useful in conditions of

muscular degeneration, skin and connective tissue regeneration and epilepsy.

Proline

- An important component of muscle and collagen (connective tissue).
- Vitamin C is essential for its incorporation into these supporting structures of the body.
- Essential for skin health.

Ornithine

- Important metabolically but not incorporated into protein.
- A very powerful stimulator of growth hormone production by the pituitary gland.
- Increases body metabolism of fat and enhances transportation of amino acids to cells.
- Involved with arginine in ammonia detoxification in the urea cycle.

Therapeutic uses
- Enhances wound healing and stimulates the immune system.
- May be useful in auto-immune diseases such as rheumatoid arthritis.

Threonine

- This essential amino acid is deficient in grains.
- Found abundantly in pulses, making a combination of grains and pulses a complete source of protein for vegetarians.
- Required for digestive and intestinal tract function.
- Prevents accumulation of fat in the liver.
- Suggested to be essential for mental health.

Therapeutic uses
- No specific therapeutic roles have been ascribed to it thus far.

Deficiency
- Irritability and personality disorders.
- Indigestion, malabsorption, malnourishment.

Genetically acquired disorders
- Hyperthreonineaemia, for which there is no adequate treatment.

CHAPTER 10

Summary of Conditions Helped by Amino Acids

Detoxification of heavy metals:
Methionine, cysteine, cystine and glutathione. (Sulphur containing amino acids).
Histidine.

Counteracting effects of free radical activity:
Methionine.
Glutathione.

Assistance in fatty metabolism:
Methionine.
Taurine.
Carnitine.

Acceleration of wound healing:
Arginine.
Proline/hydroxyproline (collagen-connective tissue regeneration).

Control of viral infection:
Lysine.

Thymus activity enhancement:
Arginine.

Glucose tolerance improvement/enhanced insulin production-utilization:
Arginine.
Taurine.

Glutamic acid.
Cysteine.
Glycine.

Immune system enhancement:
Glutathione.

Rheumatoid Arthritis:
Histidine.

Brain detoxifier (of ammonia):
Glutamic acid.

Brain detoxifier (of histamine in histadelic schizophrenia):
Methionine.

Protection against radiation effects:
Histidine.
Glutathione.
Cystine.

Weight Control – Obesity:
Phenylalanine. ⎫ Appetite control and better food
(Tryptophan). ⎭ selection.
Valine.
Methionine.
Carnitine (mobilizing fat deposits).

Depression:
Phenlyalanine.
(Tryptophan).
Tyrosine.
Glutamine/glutamic acid.

Infertility:
Arginine.
Carnitine.

Insomnia:
(Tryptophan).

Epilepsy:
Taurine.

Ageing process – skin and soft tissues:
Proline/hydroxyproline (with Vitamin C).
Glutathione.

Ageing process – general:
All amino acids.
Arginine and ornithine (promote growth hormone).
Glutathione (prevents cross linkage through free radical activity).

Cholestasis:
Taurine.

Circulatory disorders (intermittent claudication etc.):
Carnitine.
Taurine.

Concentration ('brain fuel'):
Glutamic acid (derives from glutamine).

Behavioural problems:
Glutamic acid.
Threonine (if deficient).
(Tryptophan).
Taurine.

Alcoholism and alcohol induced damage:
Glutamine.
Cystine.

Peptic ulcer:
Glutamine.

Hair health:
Cystine.

Tumours – (animal study):
Glutathione.

Lipid peroxidation deactivator:
Glutathione.

Myocardial infarction protection:
Carnitine.
Methionine.
Taurine (spares potassium).

Pain control enhancement:
(Tryptophan).
Phenylalanine (d- and l- as DLPA)
or d-Phenylalanine.

Chronic disease/nervous system degeneration:
All amino acids.
Isoleucine/leucine/valine.
Cysteine/cystine (essential for B_6 utilization).
Phenylalanine.

Parkinson's disease:
Tyrosine.

Muscular dystrophy:
Carnitine (possibly).

Drug damage (protection):
(Tryptophan). ⎫
Lysine. ⎬ animal studies
Cysteine/cystine. ⎭

Allergic conditions:
All amino acids.
Specific amino acids according to indications.

CHAPTER 11

Cautions, Combinations and Dosages

Because a nutrient, or anything else for that matter, is good for you in certain circumstances, at a particular dosage, it does not mean that you will always need it, or that more than the dosage will be even better for you.

Amino acids are no exception to this basic therapeutic rule.

As indicated in the main section of the book, in which we have looked at many of the therapeutic studies which prove the potential of amino acids to be real and exciting, the dosages vary considerably from a few hundred milligrams in some cases to several grams many times daily. It is suggested that the guidelines in that section and those listed be followed, as long as there are no contraindications (also discussed below).

Caution

Phenylalanine (eg DLPA), tyrosine and tryptophan should never be taken when drugs of the class of *monoamine oxidase (MAO) inhibitors* are being used, as this could prove dangerous.

Arginine and *ornithine* should not be used by *schizophrenics* unless under supervision and *arginine* is not advised for individuals with active *herpes infection. Ornithine may be employed instead of arginine in such cases.*

Cysteine should not be used by *diabetics*, especially when insulin is being used.

Cystine should not be used by people with a tendency towards *kidney or bladder stones*, and the use of *cysteine* should always be accompanied by three times the dosage of vitamin C to ensure that excess cystine is not produced. Always take *histidine* together with vitamin C.

Histidine should be used cautiously by anyone with a *schizophrenic* condition as some forms (histadelic) show an excess of its derivative histamine. *Manic depressives* should also avoid *histidine*. *Histidine* in dosages of 4 g daily and over can result in *onset of menstruation*. This can be used to control the timing of the menstrual flow.

Aspartic acid may result in flatulence.

Methionine use should always be together *with vitamin B_6 to avoid excess buildup of homocysteine*. It is also advisable to *always take magnesium when using methionine*. Methionine has a 'rotten egg' smell and should be used in capsule form to avoid this.

Menopausal women should take care regarding methionine supplementation.

Anyone with *high blood pressure* who is taking medication for this should be cautious with use of *phenylalanine*.

Pregnant women or those lactating should avoid use of amino acid supplementation unless under strict guidance. Any woman anticipating conception or already pregnant should avoid use of tryptophan or phenylalanine.

Tyrosine should be avoided by anyone with melanoma.

Unless specifically advised do not use 'D' forms of amino acids except in the form of pain relieving DLPA or DPA amino acids. The 'L' form is the way nature makes amino acids.

Use of individual amino acids can result in imbalances being produced among the other amino acids. This calls for a close monitoring of long term usage of any single amino acid on its own. There is no such danger when carefully constructed blends of amino acids are prescribed to meet the needs observed in amino acid profiles, or when the full complement of amino acids is taken in their free form. For these reasons use of single amino acids should be restricted in self medication situations to brief periods of real need, not exceeding a few weeks. If a condition requires more than this amount of time to assist it, then professional advice should be sought.

Use of all 20 amino acids is suggested (ideally together with appropriate nutrient factors such as minerals and vitamins) at a time separate from individual amino acid intake, to ensure that a supply of all additional, and possibly deficient, amino acids is made available to the body. It is sometimes suggested that one should take amino acids at mealtimes but it is generally better to take them away from mealtimes, 90 minutes either side of a meal. Sometimes, as in the case of the use of tryptophan as an appetite modifier, a small amount of carbohydrate assists in the uptake of the amino acid (a little sugar or a biscuit will do nicely).

Dosages

Arginine: Up to 8 g daily.

Histidine: Between 1 and 6 g daily with vitamin C.

Leucine: 10 g daily for Parkinsonism.

Lysine: 500 to 1500 mg daily in herpes cases, with a doubling of dosage when there is an active infection.

Methionine: 200 to 1000 mg daily with magnesium and vitamin B_6.

Phenylalanine: 100 to 500 mg daily.

DLPA: 750 mg three times daily for three weeks, and then double this if no pain relief noted, for a further three weeks.

DPA: 400 mg three times daily for three weeks, and then double dosage for a further three weeks, if no pain relief noted.

Proline: 500 to 1000 mg daily with vitamin C.

Threonine: 150 to 500 mg daily.

Tryptophan: 1 g before sleep (with magnesium and B_6), 2 to 3 g for pain control and depression in divided doses. Maximum dose about 6 g daily, enhanced by carbohydrate snack at time of taking. Also enhanced by taking vitamin B_3 (niacinamide) in ratio of two parts tryptophan to one part niacinamide.

Valine: 1 g daily together with phenylalanine, methionine and tryptophan. Ratio should be 3(P): 2(V): 2(M): 1(T) in weight reduction formulation, taken before meals.

Taurine: 100 to 1000 mg daily in divided doses for epilepsy, reducing to maintenance dose of 50 mg daily.

Carnitine: 1 to 3 g daily in divided doses.

Cysteine and Cystine: 1 g three times daily for a month, then twice daily, in chronic ill health, together with vitamin B_6.

Glutathione: 1 to 3 g daily.

Alanine: 200 to 600 mg daily.

Tyrosine: 2 g three times daily for depression (for two weeks). Ideal intake is 100 mg per kilo of body weight per day.

Read the studies in Chapter 7 and note the variations in dosage used. Take advice if in any doubt at all as to dosage required.

Cysteine and Cystine: 1 g three times daily for a month, then twice daily, in chronic ill health, together with vitamin B_6.

Glutathione: 1 to 3 g daily.

Alanine: 200 to 600 mg daily.

Tyrosine: 2 g three times daily for depression (for two weeks). Ideal intake is 100 mg per kilo of body weight per day.

Read the studies in Chapter 7 and note the variations in dosage used. Take advice if in any doubt at all as to dosage required.

CHAPTER 12

Amino Acid Profiles in Action (AIDS & Candida)

How we can measure amino acid imbalances

It is possible to study amino acid levels and ratios (the relationship of amino acids to each other in quantitative terms) by analysing urine, serum and other tissues. Such amino acid 'profiles' have shown definite patterns which relate to different conditions, so that it is now possible to give certain amino acids as supplements for various chronic diseases and general health problems. This is being done with great effect in cases of AIDS and chronic fungal infections such as Candida albicans. Thus amino acids can be used to supplement deficiencies, by giving one or other of these as a means of restoring normal levels or ratios, and therefore normal function.

In these cases, although part of the real cause of the problem is being taken care of, the reasons why deficiency of the substance exists still needs to be ascertained and dealt with, if possible (it may sometimes relate to genetic abnormalities which cannot be corrected). In other cases, as discussed earlier, amino acids are prescribed in order to achieve a pharmacological effect, there being no evidence of amino acid deficiency. Here the cause of the problem is not being addressed and without this happening real health, as opposed to relief of symptoms, cannot be regained.

It is obvious that if reliable use is to be made of amino acids, in therapeutic terms, some form of test is essential to help the practitioner to ascertain the needs of the patient.

Recent developments, notably in the field of high performance liquid chromatography (HPLC), have allowed the development of tests which can be routinely performed to show amino acid levels. These can then be compared with normal or reference ranges in order to assess amino acid disturbances and to allow for interpretive guidelines to be produced.

Since amino acids play such vital roles in the healthy organism and since they have such a major part to play in terms of the structure and function of the body, in both health maintenance and disease, the importance of a test which can aid in the assessment of their relative presence cannot be overestimated. Such a test may be used to discover aspects of the nutritional and metabolic status of the patient, as well as the effects of such factors as stress, trauma, other therapeutic measures (drugs etc.), as well as nutritional supplementation.

It should be recognized that for proper utilization to take place, in terms of maintenance and development of body tissues, as well as in the myriad processes of the body, the amino acids must be present in the correct ratio to each other. If ratios are inadequate, in specific amino acids, the ability for proper protein synthesis will be adversely affected.

Until the recent development of more sophisticated methods, such as HPLC, the only way in which an amino acid profile could be achieved involved cumbersome, time consuming and expensive methods, such as calorimetric and microbiological tests. HPLC is an analytical method of a degree of sensitivity, speed, reliability and relative cheapness, to allow for quantitative assessment of an individual's amino acid status.

Protein malnutrition is the frequent precursor of amino acid deficiency. Such deficiency-states may be associated with improper diet; failure to digest or absorb adequately; stress conditions; infection or trauma; drug usage; imbalances or deficiencies involving other nutrients, such as minerals or vitamins; age and its associated dysfunctions etc.

Supplementation of appropriate amino acids may result in

a restoration of normality in conditions resulting from any of these causes. It should be noted that when there exists an inadequate total calorie intake in healthy individuals, there can result a utilization of amino acids as sources of energy. This can, in turn, result in amino acid deficiency, which will not be normalized by supplementation. For this reason, when dealing with healthy individuals, the evaluation of free amino acid pools in the tissue fluids (usually plasma or urine) requires in addition the examination of the factors which might physically be affecting the energy state of the body, such as stress or the dietary pattern.[1]

As part of the routine examination, and clinical testing of the patient the addition of amino acid profiles can be seen to provide potentially useful information. Among the conditions that can thus be evaluated are inherited or secondary amino acid disorders; hepatic and renal conditions; cardiovascular conditions; disorders of the immune system; musculoskeletal problems etc. It is now becoming clear that the role of amino acid ratios, in a variety of neurological and psychiatric conditions is also of importance. A series of amino acid surveys can be used to indicate the effectiveness, or otherwise, of therapeutic measures, and of the progress, or otherwise, of disease processes. The prognostic value of such methods therefore becomes important. In their survey of the subject of amino acid analysis Dennis Meiss and John McCue[1] discuss an example of information that might be gleaned from routine analysis of this sort.

The synthesis of collagen requires vitamin C-dependent hydroxylation of certain proline and lysine residues, a manganese-dependent glycolysation of specific hydroxylysine residues, and a copper-dependent cross linking through lysyl and hydroxylysyl derivatives. The stability of collagen depends on the hydroxyproline content and on the lysine-hydroxylysyl cross-links.

Any agent or condition that interferes with amino acid uptake or utilization will have deleterious effects on the synthesis and maintenance of collagen in bone and connective tissue. Important factors contributing to this availability and utilization of amino acids involve dietary intake, balanced levels of essential and non-essential amino acids, adequate supply of cofactors for enzyme action (including vitamins and minerals) and caloric status.

Here are some examples of how amino acid analysis might detect improper or disrupted collagen synthesis. A plasma sample that is unusually high in proline and lysine might indicate inadequate conversion of these amino acids. Proline and lysine must be hydroxylated into hydroxyproline and hydroxylysine in order for collagen to have proper strength and structure.

Symptoms of defective synthesis would include pain in limbs and a general weakening of collagen in tendons and bones. The problem could be linked to vitamin C deficiency. Vitamin C is essential for hydroxylating proline and lysine: collagen synthesized in the absence of vitamin C is usually insufficiently hydroxylated and therefore less stable and easily destroyed. The recommended treatment might be vitamin C supplementation. Another analysis of proline and lysine levels at a later date could be used to monitor the effectiveness of supplementation.

A further example is given in this paper which shows another possible way in which collagen levels might be impaired, and the way in which amino acid involvement might be detected.

Unusually high levels of glycine in the plasma and urine may signal caloric or renal deficiency – a serious impediment to proper collagen formation. As in the previous instance, weakening of collagen in tendons and bones would result. When the patient is not getting enough calories, glycine in collagen may be converted into pyruvic acid for use in energy conversion (gluconeogenesis). By increasing the patient's glucose level, proper collagen-glycine levels may be restored.

The causal relationship in these two examples is reasonably clear. Other instances may not prove so clear cut, and expert interpretation of the amino acid analysis may be called for.

Currently available in the USA are amino acid profiles which, at a cost of little over one hundred dollars, provide the following:

1. A laboratory print-out showing precise levels of thirty amino acids.

2. A comprehensive discussion of these results citing up-to-date literature, for further reference.
3. Any abnormal patterns are treated in detail in that the discussion explains whether they are characteristic of (or coincident with) various pathologies or other metabolic disturbances.
4. A summary of the discussion, emphasizing probable diagnostic considerations, and where appropriate, related mineral and vitamin requirements.
5. There is also a print-out of the physiological role of each amino acid that shows as abnormal in the report.

Many US and some European laboratories who offer this service, provide collection kits for plasma and urine samples, either of which can be used for the analysis. Plasma is shipped to the laboratory packed in dry ice, which makes the possibility of transatlantic delivery an improbable procedure. The urine sample is required within forty-eight hours of collection and this may well be a possibility if arrangements were made with an airline, or with a firm specializing in rapid document delivery. The urine specimen (twenty-four hour) is collected in a special container to which has been added hydrochloric acid (provided by the laboratory as part of the kit). During the collection period this is refrigerated, and two samples of the well mixed urine-acid mixture are mailed in special containers to the laboratory. It is only a matter of time before a similar facility is available in Europe, but for the present the obstacle of the Atlantic and continental America, remains.

The amino acids reported on in an amino acid profile are as follows:[2]

Alanine; a-aminoadipic acid; a-amino-n-butyric acid; arginine; asparagine; aspartic acid; b-alanine; b-amino-isobutyric acid; citrulline; ethanolamine; glutamic acid; glutamine; glycine; histidine; homocysteic acid; homoserine; hydroxylysine; isoleucine; leucine; lysine; methionine; ornithine; phenylalanine; phosphoserine; serine; taurine; threonine; tryptophan; tyrosine; valine.

The reference ranges in the assessment and discussion, which is part of the report, represent the most statistically significant ranges which appear in the published literature on the subject. They are only guides and should not be regarded as absolute since research continues, and current beliefs may be modified by subsequent findings.

As with other tests and analysis this is not meant to be diagnostic, but to provide further evidence which, together with all other data available, should assist in the ascertainment of a diagnostic or prognostic finding.

An example from an amino acid profile report gives an idea of the assistance such a print-out might be to the practitioner:

> Histidine is low in the urine of this individual. A low urine concentration of this amino acid has been reported to accompany rheumatoid arthritis. Also histidine levels are lower in patients taking salycilates and steroids (Bremer H.J. et al., *Disturbances of Amino Acid Metabolism*, Urbasn and Schwarzenberg, 1981).
>
> Additionally a low histidine level without accompanying low levels of a majority of other essential amino acids may imply a specific dietary deficiency. Optimum metabolism of histidine is dependent upon adequate availability of folic acid.

The development of this type of nutritional profile with its potential for incorporation into the standard clinical procedures of nutritionally orientated physicians, is a major step towards the ideal of being able to predict future patterns of disease long before they manifest. It is certainly evident that disturbances in the levels of and ratios between amino acids and other nutrient factors, if corrected by nutritional manipulation and supplementation at an early stage, may well prevent degenerative states from becoming manifest. Identification of individual requirements by these means is a major part of the task of the practitioner whose aim it is to both prevent ill health and to restore it once it is lost.

Used in this way amino acid therapy is seen to be providing the body with its needs rather than addressing the symptoms of the patient and thus avoids the employment of amino acids, and other nutrients, in a pharmacological manner.

AIDS, Candida and amino acid profiles

New research with amino acid profiles has provided the first really objective test of Candida overgrowth. No fewer than seven amino acids normally not present in the urine were consistently found when a series of 70 people diagnosed by physicians as having Candida problems was tested.

When nutritionist Jeffrey Traister, MS, described the results to the American Society of Clinical Pathology in Orlando, Florida, in 1986, he also made the point that one of the amino acids that showed up, anserine, is not found in human tissue. (27 September 1986, when he presented a paper entitled 'Abnormal Amino-aciduria Pattern in Patients Diagnosed with Candidiasis'.)

The 70 people showed very distinct patterns of abnormal amino acid levels in a 24-hour urine sample. The levels in urine were compared with those of 99 patients in a reference group, who had pathology other than Candidiasis.

The following amino acids and their metabolites were found to be abnormally high. (The figure following each of these represents the percentage of the 70 patients with the abnormality):

Phosphoserine (99%); Phosphoethanolamine (33%); B-Alanine (66%); GABA (100%); Ethanolamine (66%); Hydroxylysine (100%); Ornithine (96%); 1-Methyl-Histidine (41%); 3-Methyl-Histidine (43%); Anserine (53%); Carnosine (47%); Ammonia (93%). The one amino acid found consistently below what is considered a reference range was Arginine (96%).

All the ranges, means and medians of the figures as analysed were found to be significantly higher in the Candida patients as compared to the reference group.

The theory is that Candidiasis is accompanied by

metabolic disorders involving phospholipid metabolism, liver and kidney function, maldigestion and malabsorption, connective tissue problems and altered vitamin B_6 metabolism.

Doctors in America are excited by this research, since there has been a great deal of controversy over the existence and extent of the suggested Candida overgrowth 'epidemic'. Now there is an objective way of proving Candida activity to be a part of an individual's problem. Usually this takes enlightened guesswork. Almost everyone on earth is infected by Candida and there is no easy way of knowing whether its current activity is within normal or excessive bounds. Symptoms are only a clue, since there are so many that can be caused by Candida overgrowth.

The amino acid profile is useful in definitely identifying Candida as the culprit, and in monitoring changes as therapy continues and the profiles alter back towards 'normal'.

The AIDS profile

These profiles are far easier to read than they look. The name of the substance is listed on the left (see pages 195 to 200).

The next three columns show whether it was 'low', 'in-range' or 'high' compared with the 'reference range' for adults, which is listed on the extreme right of the page.

If we look at the amino acid profile of a person with AIDS who died in August 1986 (pages 195 onwards), we can see that in the column on the left, after the name of the substance being assessed, marked 'low', a number of substances are listed. Among these are several in which the column carries the legend 'ND': none detected at all. This was the case for phosphenthanolamine, aspartic acid, glutamic acid, proline, amino-n-butyric acid, valine, cystathionine, ethanolamine, lysine, 3-methylhistidine, arginine.

Listed as low, but detectable, were threonine, asparagine and 1-methylhistidine.

A number of amino acids also appear in the list of those

which are high by comparison with normal ranges. These were: a-aminoadipic acid, cystine, methionine and phosphoserine.

In the list of amino acids which could not even be detected were a number of amino acids which are necessary for protein synthesis in the body including valine, proline, arginine and lysine, aspartic acid and glutamic acid.

Just what is the significance of such low levels in the urine in terms of the various metabolic processes in which they are involved, is a matter of ongoing debate among experts. But what is already clear is that there is a fairly consistent pattern of amino acids being found in the serum and urine of PWAs. That means that the profile can show progress, or lack of it.

If profiles of both serum and urine are taken, they will show virtually what is in the blood and what is being excreted. So some degree of manipulation of the amino acid status with appropriate supplementation can be undertaken.

A look at the profiles of the 'typical' ARC person and one HIV-positive also shows detectable abnormalities that are unique to that condition (and to some extent, of course, to the particular person).

A short study of the profiles will show the marked differences, with the AIDS picture one of a vast number of essential substances in the body being exhausted.

Why supplement?

Putting some of these elements back into the body in a form that is readily available for conversion is not the answer to AIDS, or any other immune system condition, or chronic health problem. But it does offer real assistance in a desperate situation.

There is no suggestion that this is in any way 'curative'. However, it seems possible that supplementation of this sort allows the person to continue to function while a variety of other things are being done to help restore normality; in other words a chance to stay alive while healing occurs.

Amino acids alone cannot do this. A whole team of interacting nutrients are required. What is certain is that this cannot be done without amino acid supplementation or some other method that will ensure that these essential protein fractions are provided to the body in a usable form.

(The authors wish to acknowledge the assistance given by Dr Don Tyson, of Tyson & Associates, Santa Monica, California, in preparation of this discussion on amino acids and AIDS.)

Dr Tyson makes it clear in personal communications with the authors (3 August 1987) that until his organization has documented results of survival of people with AIDS/ARC for at least three years, they are making no claims. The longest survivor currently on the programme is 2.5 years.

He states:

> We are pursuing the immune system via amino acid therapy based on laboratory amino acid analysis and other laboratory support data in putting together a programme for AIDS patients, and presently we have patients who are doing well clinically.
>
> Their laboratory data shows significant improvement as far as T and B cell ratio with overall immune function improvement. We must have survival for a minimum of three years; then and only then could I feel competent that we are doing good.

In the meantime there is patently a strong case for such an approach which can only be seen as supportive of the multiple efforts towards enhancing the health status of very sick people.

Date Reported: D.O.B.:   Physician: N.F.
Log No.: FIRST Assay   Sex: MALE   Address:
Patient: R.L.   Height:
Address: AIDS/DIED 8-'86   Weight:
Phone: Tot.Vol.:   Phone:

AMINO ACID (urine) *Ammonia range subject to change	PATIENT VALUE *ND – nondetectable			ADULT Urine Reference Range Micromol 24/hr
	LOW	In-range	HIGH	
1. Phosphoserine			163	ND
2. Taurine		406		63–2300
3. Phosphoethanolamine	ND			26–101
4. Aspartic Acid	ND			23–218
5. Hydroxyproline		ND		ND
6. Methionine sulfoxide		ND		ND
7. Threonine	60			85–440
8. Serine		407		160–700
9. Asparagine	59			270–700
10. Glutamic Acid	ND			55–270
11. Glutamine		198		140–860
12. Sarcosine		ND		ND
13. A-Aminoadipic Acid			94	31–81
14. Proline	ND			80–120
15. Glycine		515		160–4200
16. Alanine		224		60–800
17. Citrulline		ND		ND
18. A-Amino-n-butyric Acid	ND			16–77
19. Valine	ND			14–51
20. Cystine			131	20–130
21. Methionine			212	20–95
22. Cystathionine	ND			16–63
23. Isoleucine		71		18–210
24. Leucine		91		21–200
25. Tyrosine		94		40–270
26. Phenylalanine		98		24–190
27. B-Alanine		ND		ND
28. B-Aminoisobutric Acid		71		0–500
29. Homocystine		ND		ND
30. G-Aminobutyric Acid		ND		ND
31. Tryptophan		ND		ND
32. Ethanolamine	ND			59–307
33. Hydroxylysine		ND		ND
34. Ornithine		ND		ND
35. Lysine	ND			48–640
36. 1-Methylhistidine	37			130–930
37. Histidine		318		130–2100
38. 3-Methylhistidine	ND			56–249
39. Anserine		ND		ND

AMINO ACID (urine) *Ammonia range subject to change	PATIENT VALUE *ND – nondetectable			ADULT Urine Reference Range Micromol 24/hr
	LOW	In-range	HIGH	
40. Carnosine		ND		ND
41. Arginine	ND			64–96
42. Ammonia		3791		0–5000

Aatron Medical Services
1661 Lincoln Boulevard Suite 300
Santa Monica, Ca 90404

CA Lab.Lic. No. 01451
Insurance RVS No. 82130
Phone (213) 452-7881

Harry M. Bauer, M.D.
Medical Director
CA License No. G3227

Date Reported: 10-17-86   D.O.B.:   Physician:
Log No.: 6544.U   Sex: F   Address:
Patient: ARC   Height:
Address: Weight:
Phone: Tot.Vol.: 1410cc Phone:

AMINO ACID (urine) *Ammonia range subject to change	PATIENT VALUE *ND – nondetectable			ADULT Urine Reference Range Micromol 24/hr
	LOW	In-range	HIGH	
1. Phosphoserine			128.8	ND
2. Taurine			7126.0	63–2300
3. Phosphoethanolamine			125.8	26–101
4. Aspartic Acid		100.4		23–218
5. Hydroxyproline		ND		ND
6. Methionine sulfoxide		ND		ND
7. Threonine	84.0			85–440
8. Serine		283.5		160–700
9. Asparagine	149.2			270–700
10. Glutamic Acid	70.5			55–270
11. Glutamine		457.8		140–860
12. Sarcosine		ND		ND
13. A-Aminoadipic Acid			124.6	31–81
14. Proline	ND			80–120
15. Glycine		1118.6		160–4200
16. Alanine		400.2		60–800
17. Citrulline		ND		ND
18. A-Amino-n-butyric Acid		35.2		16–77
19. Valine		26.3		14–51
20. Cystine		28.0		20–130
21. Methionine			114.8	20–95
22. Cystathionine		20.2		16–63
23. Isoleucine		25.0		18–210
24. Leucine		50.4		21–200
25. Tyrosine	39.2			40–270
26. Phenylalanine		60.2		24–190

AMINO ACID (urine) *Ammonia range subject to change	PATIENT VALUE *ND – nondetectable			ADULT Urine Reference Range Micromol 24/hr
	LOW	In-range	HIGH	
27. B-Alanine			306.6	ND
28. B-Aminoisobutric Acid		378.0		0–500
29. Homocystine		ND		ND
30. G-Aminobutyric Acid			103.2	ND
31. Tryptophan		ND		ND
32. Ethanolamine			536.2	59–307
33. Hydroxylysine		ND		ND
34. Ornithine			29.4	ND
35. Lysine	32.2			48–640
36. 1-Methylhistidine			7273.0	130–930
37. Histidine		700.4		130–2100
38. 3-Methylhistidine	177.8			56–249
39. Anserine			238.0	ND
40. Carnosine			110.6	ND
41. Arginine	TRACE			64–96
42. Ammonia			22,724.8	0–5000

Aatron Medical Services
1661 Lincoln Boulevard Suite 300
Santa Monica, Ca 90404

CA Lab.Lic. No. 01451
Insurance RVS No. 82130
Phone (213) 452-7881

Harry M. Bauer, M.D.
Medical Director
CA License No. G3227

Date Reported:
Log No.: 2nd Assay
Patient: A.G.
Address: AIDS CARRIER
Phone:

D.O.B.:
Sex: FEMALE
Height:
Weight:
Tot.Vol.:

Physician: D.W.
Address:

Phone:

AMINO ACID (urine) *Ammonia range subject to change	PATIENT VALUE *ND – nondetectable			ADULT Urine Reference Range Micromol 24/hr
	LOW	In-range	HIGH	
1. Phosphoserine			599	ND
2. Taurine		1415		63–2300
3. Phosphoethanolamine			370	26–101
4. Aspartic Acid		46		23–218
5. Hydroxyproline		ND		ND
6. Methionine sulfoxide		ND		ND
7. Threonine		184		85–440
8. Serine		302		160–700
9. Asparagine	ND			270–700
10. Glutamic Acid		102		55–270
11. Glutamine		182		140–860
12. Sarcosine		ND		ND
13. A-Aminoadipic Acid	ND			31–81

AMINO ACID (urine) *Ammonia range subject to change	PATIENT VALUE *ND – nondetectable			ADULT Urine Reference Range Micromol 24/hr
	LOW	In-range	HIGH	
14. Proline	ND			80–120
15. Glycine		1117		160–4200
16. Alanine	29			60–800
17. Citrulline		ND		ND
18. A-Amino-n-butyric Acid	ND			16–77
19. Valine	ND			14–51
20. Cystine		34		20–130
21. Methionine		74		20–95
22. Cystathionine	2			16–63
23. Isoleucine	ND			18–210
24. Leucine	ND			21–200
25. Tyrosine		74		40–270
26. Phenylalanine		37		24–190
27. B-Alanine			199	ND
28. B-Aminoisobutric Acid		125		0–500
29. Homocystine		ND		ND
30. G-Aminobutyric Acid			752	ND
31. Tryptophan		ND		ND
32. Ethanolamine		80		59–307
33. Hydroxylysine		ND		ND
34. Ornithine			171	ND
35. Lysine		65		48–640
36. 1-Methylhistidine			13,198	130–930
37. Histidine		137		130–2100
38. 3-Methylhistidine		154		56–249
39. Anserine			405	ND
40. Carnosine		ND		ND
41. Arginine	ND			64–96
42. Ammonia			50,793	0–5000

Aatron Medical Services
1661 Lincoln Boulevard Suite 300
Santa Monica, Ca 90404

CA Lab.Lic. No. 01451
Insurance RVS No. 82130
Phone (213) 452-7881

Harry M. Bauer, M.D.
Medical Director
CA License No. G3227

Log No.: SECOND ASSAY D.O.B.:   Physician: D.W.
Patient: A.G.   Sex: F   Address:
Address: AIDS CARRIER   Height:
                 Weight:

AMINO ACID (urine) *Ammonia range subject to change	PATIENT VALUE *ND – nondetectable			ADULT Urine Reference Range
	LOW	In-range	HIGH	Micromol 24/hr
1. Phosphoserine			0.6	0–trace
2. Taurine	3.3			3.5–14
3. Phosphoethanolamine		TRACE		trace–0.3
4. Aspartic Acid	0.2			1.1–5.4
5. Hydroxyproline			0.2	ND
6. Methionine sulfoxide		ND		ND
7. Threonine	5.5			7.5–25
8. Serine		6.9		6.1–19
9. Asparagine		3.6		2.6–8.6
10. Glutamic Acid		4.8		0–12
11. Glutamine		36.0		42–76
12. Sarcosine		ND		ND
13. A-Aminoadipic Acid		ND		ND
14. Proline	4.3			8.9–44
15. Glycine		15.5		13–49
16. Alanine		25.5		17–50
17. Citrulline	1.0			1.2–5.5
18. A-Amino-n-butyric Acid		1.2		0.8–3.5
19. Valine			33.7	12–33
20. Cystine	2.8			3.1–14
21. Methionine		3.7		1.3–3.9
22. Cystathionine		ND		ND
23. Isoleucine	2.4			3.5–10
24. Leucine	3.4			6.9–16
25. Tyrosine	2.4			3.2–8.7
26. Phenylalanine	3.0			3.4–12
27. B-Alanine		ND		ND
28. B-Aminoisobutric Acid		ND		ND
29. Homocystine		ND		ND
30. G-Aminobutyric Acid			2.1	ND
31. Tryptophan	1.6			1.9–7.3
32. Ethanolamine			1.9	trace–1.1
33. Hydroxylysine		ND		ND
34. Ornithine	2.2			3.0–13
35. Lysine	5.7			9.0–26
36. 1-Methylhistidine		ND		ND
37. Histidine	2.2			5.6–12
38. 3-Methylhistidine		ND		ND
39. Anserine		ND		ND
40. Carnosine		ND		ND

AMINO ACID (urine) *Ammonia range subject to change	PATIENT VALUE *ND – nondetectable			ADULT Urine Reference Range Micromol 24/hr
	LOW	In-range	HIGH	
41. Arginine	2.2			4.6–15
42. Ammonia			12.2	0–5

Aatron Medical Services
1661 Lincoln Boulevard Suite 300
Santa Monica, Ca 90404

CA Lab.Lic. No. 01451
Insurance RVS No. 82130
Phone (213) 452-7881

Harry M. Bauer, M.D.
Medical Director
CA License No. G3227

Essential Amino Acid Content of Common Foods

(Derived from *Nutrition Almanac*, McGraw Hill, 1979)

Food	Weight (gms)	Protein (gms)	TRP (mg)	LEU (mg)	LYS (mg)	MET (mg)	PHA (mg)	ISL (mg)	VAL (mg)	THR (mg)
Wholewheat bread	23	2.1	29	166	71	37	117	106	113	72
Wholewheat flour	120	15	192	1072	432	240	784	688	739	464
Soya flour	110	45	605	3428	2784	650	2179	2380	2339	1734
Oatmeal (cooked)	236	4.7	76	501	221	86	275	275	319	205
Brown rice (raw)	190	14.3	159	1233	558	260	717	675	1004	558
Brown rice (cooked)	150	3.8	41	327	148	68	190	179	266	148
White rice (cooked)	150	3	33	258	117	54	150	141	210	117
Wheatgerm	6	1.8	16	110	99	26	58	76	88	86
Cottage cheese	260	44.2	469	4608	3584	1195	2304	2475	2475	2005
Edam cheese	28	7.7	108	775	591	211	429	523	575	300
Parmesan cheese	28	10	140	980	730	260	540	670	720	370
Egg-boiled/raw	50	6.5	102	559	406	197	369	420	470	318
Buttermilk	246	8.9	90	809	678	188	433	515	613	384
Skim milk	64	23	320	2220	1780	570	1095	1461	1575	1073
Yogurt (part skim)	250	4.3	93	842	706	196	450	536	638	400

Food	Weight (gms)	Protein (gms)	TRP (mg)	LEU (mg)	LYS (mg)	MET (mg)	PHA (mg)	ISL (mg)	VAL (mg)	THR (mg)
Fish, Cod (canned)	453	87	870	6609	7655	2523	3216	4435	4611	3742
Shrimp (cooked)	453	92	821	6240	7225	2381	3038	4187	4351	3530
Trout (raw)	453	97	974	7302	8571	2827	3606	6654	6930	5621
Orange	180	1.8	5	–	48	5	–	–	–	–
Peach	100	0.68	4	29	30	31	18	13	40	27
Strawberries	149	1.04	13	63	48	1.5	34	27	34	37
Beef (roast)	453	108	1154	7888	8369	2405	3944	5002	5291	4233
Liver (cooked)	453	120	1354	8398	6772	2167	4515	3786	5689	4334
Lamb	453	80	1525	9075	9543	2884	4832	6131	5768	5421
Chicken (breasts)	358	74.5	894	5438	6630	1937	2980	3948	3800	3204
Almonds	133	25	234	1934	774	344	1524	1161	1495	811
Brazils	167	23	312	1885	740	1571	1030	990	1374	705
Peanuts (roasted)	240	60	800	4432	2592	640	3680	2992	3616	1952
Pumpkin seeds	230	67	1201	5269	3068	1267	3735	3735	3602	2001
Sesame seeds	230	42	711	3461	1256	1382	3181	2052	1925	1548
Walnuts	100	16	175	1228	441	306	767	767	974	589
Lima beans (raw)	100	20	202	1628	1488	250	1212	992	1030	836
Beans (green-cooked)	125	2	28	116	104	30	48	90	96	76
Carrots (cooked)	150	1.35	11	77	62	11	50	54	66	51
Chickpeas dry-raw (garbanzos)	100	20.5	164	1517	1415	266	1004	1189	1004	738
Lentils (cooked)	100	15.6	140	954	898	100	654	540	626	496
Mushrooms (tinned)	200	3.8	12	444	–	266	–	840	596	–
Potato (baked)	100	2.6	26	130	138	31	114	114	138	107

Food	Weight (gms)	Protein (gms)	TRP (mg)	LEU (mg)	LYS (mg)	MET (mg)	PHA (mg)	ISL (mg)	VAL (mg)	THR (mg)
Soybeans (cooked)	200	22	330	1870	1518	330	1188	1298	1276	846
Tomato (raw)	150	1.65	15	68	69	12	46	48	46	54

Index